Reasons of State

A volume in the series

Cornell Studies in Political Economy

EDITED BY PETER J. KATZENSTEIN

A full list of titles in the series appears at the end of the book

Reasons of State

Oil Politics and the Capacities of American Government

G. John Ikenberry

Cornell University Press

Ithaca and London

First published 1988 by Cornell University Press.

International Standard Book Number (cloth) 0-8014-2155-1
International Standard Book Number (paper) 0-8014-9488-5
Library of Congress Catalog Number 88-3660
Printed in the United States of America
*Librarians: Library of Congress cataloging information
appears on the last page of the book.*

*The paper in this book is acid-free and meets the guidelines for
permanence and durability of the Committee on Production Guidelines
for Book Longevity of the Council on Library Resources.*

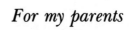

For my parents

Contents

Preface

What capacities does the modern American state possess to cope with rapid international political and economic change? This is the core question informing this book, and it emerges from recent historical events and theoretical debates. Events of the last fifteen years underscore the importance of national adaptation. International economic and political hierarchies are again changing as profoundly as they did in the 1870s and 1940s. Amid these shifting historical sands, the international power of the United States is eroding. In the decades before World War I, the United States could isolate itself from international political and economic change. After World War II the unprecedented power of the United States allowed it to set the terms of adjustment and change. But the last quarter of the twentieth century presents new realities; the United States is both more involved in the global political economy and less able to shape its course.

Theorists of international relations and political economy have taken note of these historical developments, of course, and a variety of theoretical debates have raged over the relationships among the power of the state, political sovereignty, and economic interdependence. Since the early 1970s, old questions of order and change in international relations have been examined anew. Traditional conceptions of the state and the state system have been challenged by scholars embracing a new interest in the multinational corporation, transnational organizations, bureaucratic politics, and transgovernmental relations. To some the changes were ephemeral; to others, profound.

To address these large-scale theoretical problems, I focus on the issue of "state capacity," and do so by investigating a key set of historical events that propelled these controversies—the oil shocks of the 1970s. In examining the capacity or "weightiness" of the state, particularly the American state, I look to understand the constraints on and opportunities for states—as organizational entities—to realize the goals they embrace. At moments of crisis, what can states accomplish? What international and national resources and instruments of power can state officials marshal in the face of international change? How do domestic structures provide opportunities or set limits on the ability of states to achieve official objectives?

In the United States, I argue in this book, policy responses to the oil shocks were shaped and channeled by the institutional structures of the state. Various policy responses were attempted, but those which succeeded were the responses that built upon existing channels of state action. The circuitous path of American energy adjustment since 1973 is explicable in terms of institutional configurations of the state which predated the first oil shock. The historical legacy of earlier policy struggles and the broad sweep of state building are essential guides if we are to understand the capacity of the American state in the 1970s and beyond. As a consequence, it is important to probe the historically unique and institutionally circumscribed character of state intervention in the economy and the society.

The deep significance of these institutional forces is underlined by two additional themes that emerge in this book. First, the opportunities and constraints that institutional structures mold for state action influence the venues in which state officials seek to solve adjustment problems. In particular, domestic institutional constraints on American energy policy initially encouraged policy makers to resort to international strategies of adjustment. At the same time, however, the absence of similar constraints in other industrial oil-consuming nations, together with the inability of American officials to live up to their foreign commitments, thwarted international agreements to contain the effects of rising oil prices. The interlocking nature of domestic and international energy politics is thus revealed.

Second, the eventual American resort to market pricing in order to achieve energy adjustment reflects the importance of the market as an instrument of state capacity. The extension or maintenance of markets can be a powerful tool of the state, a tool used in the service of national and geopolitical objectives. In a nation such as the United States, where it is difficult to build visible new institutional capacities for the state, manipulation of the market has great importance as an instrument of national policy.

The theoretical task at hand is to understand the role and dynamics of institutional change and the manner in which institutional structures shape, bend, alter, and blunt ongoing policy struggles. The theoretical enterprise does not end with this book. The goal, an adequate theory of institutions, remains as important as it is elusive. At this stage we need malleable theoretical templates that are, as Susanne Hoeber Rudolph suggests in an evocative metaphor, "made of soft clay rather than hard steel, that adapt to the variety of evidence and break when they do not fit." We have learned a lot, but there is still a lot to know.

The overall argument of this book has not been published elsewhere, although parts of it have been presented in various articles. The typologies of adjustment strategies presented in Chapter 1 contain materials previously published in "The State and International Strategies of Adjustment," *World Politics*, October 1986. This material is adapted by permission of Princeton University Press. "The Irony of State Strength: Comparative Responses to the Oil Shocks," published in *International Organization*, Winter 1986, contains some of the core ideas on state capacity which appear in Chapter 8. Sections of this article are used, in modified form, by permission of the MIT Press. Chapter 7 contains materials that also appear in "Market Solutions for State Problems: The International and Domestic Politics of American Oil Decontrol," *International Organization*, Winter 1988. Chapters 2 and 3 contain materials that also appear in "Conclusion: An Institutional Approach to American Foreign Economic Policy," *International Organization*, Winter 1988.

This book has benefited from the generous assistance of teachers, friends, and colleagues. I owe special thanks to Charles Lipson and Theda Skocpol, my teachers at the University of Chicago. Each provided intellectual guidance and encouragement over many years and in ways that will be difficult to repay. Theda Skocpol was particularly helpful as I wrote this book. As teachers, both have given me standards of scholarly excellence that will continue to inspire me throughout my professional career. Lloyd Rudolph was also an important adviser in the early stages of this project. I also owe a special debt of gratitude to David A. Lake and Michael Mastanduno, who have touched this book in what to them may have seemed endless sessions of discussion and criticism. Equally important was their camaraderie, which began five years ago when we were Research Fellows at the Brookings Institution. These personal and professional relationships have enriched this project in untold ways.

A score of other friends and colleagues have improved this book with their advice and criticism. Beverly Crawford, I. M. Destler, Jeff Frieden, Barbara Geddes, Judith Goldstein, Peter Gourevitch, Joanne Gowa, Stephan Haggard, Peter Hall, Jeffrey Hart, Miles Kahler, Robert Keohane, Stephen Krasner, John Odell, Ken Oye, Robert Putnam, Richard Samuels, Duncan Snidal, and Peter VanDoren provided criticisms and suggestions on chapters of the book or on articles that preceded it. Peter Cowhey, Harvey Feigenbaum, Peter Katzenstein, and Edward Morse read an earlier draft of the entire manuscript and provided detailed suggestions. Tom Ryan and David Pavelchek provided valuable research assistance. Valerie Kanka provided careful secretarial and research assistance. The editorial efforts of Roger Haydon at Cornell University Press measurably improved the manuscript.

As is appropriate for a book that stresses the importance of institutions, I have received considerable support from several organizations. Early support for the project was provided by the Brookings Institution, where I was a Research Fellow in the Foreign Policy Studies Program during 1982–83. Writing was also carried on at the University of Chicago with financial assistance from the Institute for the Study of World Politics and the Program on Interdependent Political Economy. At Princeton University, the Woodrow Wilson School for Public and International Affairs generously provided summer funding and other forms of support. Travel and research assistance was also funded by the university's Committee on Research in the Humanities and Social Sciences. The final preparation of the manuscript was made possible by financial assistance from the Center for International Studies and its director, Henry Bienen. The Institute for Advanced Study provided a refuge for the final preparation of the book.

This book is dedicated to my parents, Nelda B. Ikenberry and Gilford J. Ikenberry, Jr., who across the years have provided love and moral support without fail—the rarest and most important ingredient of all.

G. John Ikenberry

Princeton, New Jersey

Reasons of State

CHAPTER ONE

The Oil Shocks and State Responses

> The embargo on oil by the oil-producing states in the winter of 1973–74 together with the drastic rise in the price of oil have clarified like a stroke of lightning certain basic aspects of world politics which we might have understood theoretically simply by reflection but which were brought home by the drastic change in power relations which these two events imply.
>
> Hans J. Morgenthau, 1975

The political and economic foundations of the Western world were shaken in the 1970s by an upheaval in international petroleum markets. A group of small, oil-producing nations in the Middle East engineered two momentous price shocks, in 1973–74 and 1979. To many observers at the time the rise of the Organization of Petroleum Exporting Countries, and the economic turmoil that surrounded it, marked a turning point in postwar history. The remarkable postwar expansion of the advanced industrial economies was at an end; no longer could governments promise unlimited economic growth. Nor could they retain exclusive control over the management of the world economy; developing countries, particularly those rich in resources, had to be brought into the system. Most important, American postwar leadership, already perceived to be on the wane, looked to have been dealt another, perhaps decisive, blow. An era was ending. The shape of the new one remained to be negotiated.[1]

[1]One contemporary historian, writing in 1975, argued that the economic and political turmoil generated by higher oil prices was "the opening stage of a struggle for a new world order, a search for positions of strength in a global realignment, in which the weapons (backed, naturally, by the ultimate sanction of force) are food and fuel." Geoffrey Barraclough, "Wealth and Power: The Politics of Food and Fuel," *New York Review of Books*, August 7, 1975, p. 23.

From the perspective of a later decade it seems there was less change than met the eye. The industrial nations weathered the oil price shocks, even if particular political leaders did not. The political and economic power of OPEC has come and gone. The prospect of far-reaching negotiations to advance the economic plight of developing nations has never appeared more fanciful.

Nonetheless, the oil price upheavals of the 1970s generated challenges that were real enough. They signaled not just higher energy prices but also a broader socioeconomic crisis, on a scale not seen for a generation, as they confronted the countries of the industrial world with a variety of dilemmas. The oil shocks cast up insidious problems—issues of energy security, economic adjustment, and leadership and cooperation within the industrial world. The international dilemmas were variously defined, and a host of policy responses were brought to bear on them. Cooperative international responses aimed at mitigating the severity and effects of price increases; they largely failed. Across the industrial world, government leaders pursued separate national policies, some more successfully than others.

The United States, though less dependent on imported OPEC oil than other industrial nations, was no less pressed to decide how to adjust its economy and society to new international energy markets. Higher energy costs engaged officials responsible for both foreign policy and domestic policy, and proposals for action found their way onto many different policy agendas. Across political and economic life, very little was beyond the reach of the oil price revolution.

International crises of this sort pose intriguing questions for students of politics. Crises are, as one analyst notes, "critical moments when national character and institutions [are] thought to be decisively tested."[2] I share the conviction that political responses to crises can reveal essential characteristics of a nation's institutions—their capabilities and their limitations.

In this book I am interested in two particular issues. One is the capabilities of nations and limitations on their cooperation to address the common dilemmas of the oil shocks. The possibilities for cooperation among the advanced industrial nations were obvious. Why did cooperation among oil-importing industrial nations prove so elusive? The other issue is the capabilities of politicians and executive officials, particularly in the United States, to direct the course of energy adjustment. What resources and mechanisms did American political leaders have at their disposal? In the case of American energy adjustment,

[2]Randolf Starn, "Historians and Crisis," *Past and Present* 52 (August 1971), 9.

why were so many different courses of action pursued, and why did some fail and others succeed? As I trace the circuitous path of American adjustment, both international and domestic, I answer these questions, and my answers emphasize the shaping, constraining features of national political institutions and the organizational structure of policy making.

THE OIL PRICE REVOLUTION

October 17, 1973, the day of the Arab OPEC embargo, was "energy Pearl Harbor day."[3] In response to the outbreak of war between Israel, Egypt, and Syria, the Arab members of OPEC announced an embargo on oil shipments to countries supporting Israel. The United States, which was in the midst of approving emergency military assistance to Israel, was their primary target.[4] OPEC members agreed to cut back oil production, and in the next two months world supplies dropped by about 9.8 percent. The price of oil traded on the spot market in Rotterdam rose from $6.71 a barrel in October to over $19 a barrel in December, and across 1974 the official price of OPEC oil rose from its pre-embargo level of about $3 a barrel to over $11 a barrel. A relatively stable petroleum regime, managed by the large oil firms and protected by American diplomatic and military strength, had collapsed.

In 1979, after several years of relative stability, prices again rose sharply. The second oil crisis began in late 1978 with a disruption in Iranian production sparked by domestic political upheaval. A short-term cutback in production by Saudi Arabia in January 1979 helped reduce world supply on a scale similar to 1973, and as before, prices moved rapidly higher. In the early months of 1979 spot market prices rose from $19 a barrel to $31 a barrel. By mid-1980 the weighted average of OPEC oil had moved to about $32 per barrel. In seven years crude oil prices, adjusted for inflation, had increased more than 500 percent.[5]

OPEC had dramatically reversed the terms of trade, and a massive transfer of wealth resulted. For oil-importing nations, price increases

[3]S. David Freeman, *Energy: The New Era* (New York: Vintage, 1974), p. 1.

[4]See Federal Energy Administration, Office of International Affairs, *U.S. Oil Companies and the Arab Oil Embargo: The International Allocation of Constricted Supplies*, Report prepared for Subcommittee on Multinational Corporations of the Senate Committee on Foreign Relations, 94th Cong., 1st sess., January 25, 1975, app. 1.

[5]For changes in oil prices see Philip K. Verleger, Jr., *Oil Markets in Turmoil: An Economic Analysis* (Cambridge, Mass.: Ballinger, 1982), pp. 29–53.

3

Table 1. Current account balances of OPEC, OECD, and non-oil developing
countries, 1973-1980 ($ billions)

	1973	1974	1975	1976	1977	1978	1979	1980
OPEC	9	62	30	39	31.5	5	70	120
OECD	19	−16.5	12	−5.5	−10	28	−13	−47
Non-oil developing countries	−11.5	−31.5	−37	−24.5	−19.5	−30.5	−47	−62

Current account balances are surpluses or deficits on goods, services, and private
transfers.
SOURCE. OECD, *Economic Outlook,* various issues.

raised import costs and lowered real income. The oil revenues of the
OPEC countries rose approximately $70 billion in 1974 alone, and the
transfer of real income to OPEC countries amounted to about 2 percent of the gross national product of the member-states of the Organization for Economic Cooperation and Development. During the
second oil price shock OPEC oil revenues rose about $180 billion. The
transfer of income was again about 2 percent of OECD gross national
product.[6] Data on current-account balances between OECD and
OPEC countries reflect in part the magnitude of the income transfers
involved (see Table 1).

These oil crises were high drama at the time, but signs of an impending change in relations between oil producers and consumers
appeared long before 1973. Behind the immediate actions taken by
OPEC producers was a convergence of incremental changes in the
international production and consumption of energy. Three trends
were particularly important. First, a remarkable surge in economic
growth in the mid- and late 1960s was shadowed by steady growth in
energy demand across the industrial world. Second, the production
costs of oil were very cheap, and oil gradually replaced coal as the
preeminent source of energy. Finally, while other regions of oil production were nearing capacity, the Middle East and North Africa had
almost limitless petroleum reserves.[7]

As oil became increasingly important to the Western economies and
its production became geographically more concentrated, changes
were occurring in relations between oil-producing governments and
multinational oil companies. In the late 1960s and early 1970s a large
number of independent oil companies began operations in regions

[6]See Stanley W. Black, "Learning from Adversity: Policy Responses to Two Oil
Shocks," *Essays in International Finance* no. 160 (Princeton, N.J., December 1985), p. 5.
[7]Joel Darmstadter and Hans H. Lansberg, "The Economic Background," in Raymond Vernon, ed., *The Oil Crisis: In Perspective, Daedalus* 104 (Fall 1975), 16.

exclusively controlled and managed by the major firms. More companies began to compete for oil contracts, giving the producing nations new advantages in bargaining.[8] By 1970, moreover, the posted price of Middle Eastern oil had declined every year in real terms since 1947 (except for several years in the mid-1950s).[9] Host governments had reason to renegotiate prices with the oil companies.

In 1970 Libya demanded a new concessionary agreement with the major oil companies and, by threatening nationalization and cutbacks in production, extracted new price and tax arrangements. Standing behind Libya was OPEC. In 1968 OPEC had foreshadowed this new bargaining campaign, resolving that oil prices should keep pace with the prices of industrial products and that its members should determine prices and profit shares. Libya's success and the other renegotiations that followed stemmed from the changing patterns of supply and demand noted above, together with fear of energy shortages in Europe, OPEC's newfound resolve, and the willingness of independent petroleum companies to agree to terms previously unacceptable to the major firms.[10]

Price increases would be the chief legacy of the October embargo. The attempt by Arab OPEC members selectively to embargo oil supplies failed, in large part because of the role the international oil companies continued to play in supply and distribution. Nonetheless, the OPEC challenge was perceived not just as an effort to redress longstanding economic grievances but as an attempt to alter the foreign policy of consumer nations. The threat was manifestly political as well as economic.

ENERGY ADJUSTMENT PROBLEMS

These shocks confronted Western importing countries with immediate demands to adjust their economies and address matters of energy security. Regarding issues of security of access, Western Europe and Japan were more threatened than the United States, even though the 1973 embargo was directed primarily at America. Almost all of Japan's petroleum came from foreign (mostly Middle Eastern) sources, and almost 70 percent of the energy Japan required came in the form of petroleum. France also was highly dependent on OPEC

[8]See Raymond Vernon, "An Interpretation," in ibid., pp. 1–14.
[9]Louis Turner, *Oil Companies in the International System* (London: Allen & Unwin, 1980), p. 125.
[10]Edith Penrose, "The Development of Crisis," in Vernon, *The Oil Crisis*, pp. 40–42.

oil; West Germany, with a substantial national coal industry, was less so.

Though their resources differed, the industrial importing nations faced a common set of international challenges. Most important, the industrial oil-consuming nations had a common interest in reducing demand in the international oil market. The disruptions of 1973–74 and 1979 involved only modest shortfalls in oil supplies, but both times the dramatic jump in oil prices, on the spot market and in the official prices posted by OPEC, resulted in part from the competitive scramble by users to protect themselves against loss of supply. Short-term incentives to hedge against the uncertainty of future supplies led governments and oil companies to bid up prices.

That it was a politically inspired embargo which triggered the first oil shock heightened the common Western perception that the immediate threat was to energy security. Yet economic adjustment to higher oil prices caused problems that were both broader and more profound, interacting with other long-term and cyclical problems and blurring boundaries between energy concerns and larger economic predicaments. As one study noted in 1976, "Increasing trade in petroleum and higher prices have given rise to complex economic transactions, providing every state with new opportunities for influencing other states, while making each more sensitive to the actions of others."[11] The price increases disrupted and disequilibrated a range of international markets and national economic systems. The challenges for governments involved adjusting national economies to higher energy prices and changing competititve positions.

The problems at this level were enormous. They included paradoxical macroeconomic problems—the price increases had *both* inflationary and deflationary effects on domestic economies. In the terms of traditional monetary and fiscal policy, as a result, the proper response was neither obvious nor straightforward. Traditional policy choices could address either inflation or unemployment, but they could not address both at once.[12]

The high price of OPEC oil was also reflected in balance-of-payments deficits. Immediately after the embargo, government officials

[11]Nazli Choucri, *International Politics of Energy Interdependence: The Case of Petroleum* (Lexington, Mass.: Lexington Books, 1976), p. 53.

[12]See Robert S. Pindyck and Julio J. Rotemberg, "Energy Shocks and the Macroeconomy," in Alvin L. Alm and Robert J. Weiner, eds., *Oil Shock: Policy Response and Implementation* (Cambridge, Mass.: Ballinger, 1984), pp. 97–112. See the general discussion of the macroeconomic effects of the energy crises in Peter F. Cowhey, *The Problems of Plenty: Energy Policy and International Politics* (Berkeley: University of California Press, 1985), pp. 247–50.

from all the major importing nations worried about the ability of particular nations to pay for higher-priced oil. Discussions began concerning emergency lending to nations in chronic deficit. American officials proposed a $25 billion Financial Solidarity Fund to supplement existing international monetary reserves for this particular form of adjustment. The problem of coping with such deficits also opened up a variety of subsidiary policy options.[13] Payments deficits could be attacked, for example, by selling more products abroad or by restricting oil imports. In any case, problems of industrial competitiveness quickly became a part of the dilemma.[14]

GOVERNMENT RESPONSES TO THE OIL SHOCKS

Across Europe and in Japan and the United States, national patterns can be detected in the attempts of government leaders to cope with the surge in oil prices. The divergent policies reflected differences in government priorities—whether states wanted macroeconomic adjustment and industrial competitiveness or security of energy supply, or had larger foreign policy objectives. Different emphases in turn reflected differences in objective economic circumstances, as well as differences in the policy instruments available. In the midst of a common shock, distinctive national policies emerged.

The French government emphasized national control over energy production and supply. This decision led to two initiatives. First, the French government addressed security of supply with a series of commercial agreements with OPEC producers, particularly the Saudi government. This initiative involved government-to-government, long-term commercial and barter contracts. Second, after 1973 the French government accelerated its already large commitment to civilian nuclear energy, embarking on the most ambitious nuclear power pro-

---

[13]See Jan Tumlir, "Oil Payments and Oil Debt in the World Economy," *Lloyds Bank Review*, July 1974, pp. 1–14, and Thomas O. Enders, "The Role of Financial Mechanisms in the Overall Oil Strategy," *Department of State Bulletin*, March 10, 1975, pp. 312–17. Efforts by the Interim Committee of the IMF are discussed in Robert Solomon, *The International Monetary System, 1945–1976* (New York: Harper & Row, 1977), Chap. 17. On subsidiary options see Choucri, *International Politics*, pp. 55–58.

[14]The literature on the economic impact of the oil shocks is vast. See, e.g., T. M. Rybczynski, ed., *The Economics of the Oil Crisis* (New York: Holmes & Meier, 1976); Edward Fried and Charles Schultze, eds., *Higher Oil Prices and the World Economy: The Adjustment Problem* (Washington, D.C.: Brookings, 1975); and Daniel Yergin, "Crisis and Adjustment: An Overview," in Yergin and Martin Hillenbrand, eds., *Global Insecurity: A Strategy for Energy and Economic Renewal* (Boston: Houghton Mifflin, 1982), pp. 1–28.

gram in Europe.[15] By the end of the decade France was far more reliant than the United States, West Germany, and Japan on nuclear energy. In both respects the state was at the center of the French response.

Japan and West Germany emphasized the problems of industrial competitiveness and macroeconomic adjustment. Their policies were not confined to petroleum production but addressed industrial efficiency across sectors. Both countries launched a commercial offensive, attempting to balance payments by selling more exports, particularly to the oil-exporting countries of the Middle East. They also engaged in some industrial restructuring. Japan scaled down or phased out parts of such energy-intensive industries as aluminum and petrochemicals. Both countries also launched programs that would prove to be as important as actual restructuring, to encourage industrial energy efficiency.

The Japanese strategy was the product of specific steps to address the problems of inflation, balance-of-payments deficits, growth, and unemployment. From 1973 to 1975 the government pursued a tight monetary policy to address inflationary pressures. Through 1977 exports began to restore equilibrium to trade payments—helped by a depreciated currency and overall growth in world markets. In 1978 and 1979 normal levels of growth were restored. Adjustment entailed numerous programs for energy conservation and efficiency: research on new conservation technologies, tax incentives and low-interest loans, and conservation regulations all became part of the Japanese response.[16]

The West German response was neither as elegantly formulated nor as coherently implemented. In many respects it involved an intensification of established economic policy, which worked to encourage

[15]On security of supply see Louis Turner, "Politics of the Energy Crisis," *International Affairs* (London) 50 (July 1974), 404–15, and Horst Mendershausen, *Coping with the Oil Crisis* (Baltimore: Johns Hopkins University Press, 1976). On the general rise of oil trading by states see Aliro A. Parra, "The International Role and Commercial Politics of National Oil Companies," *OPEC Review* 6 (Spring 1982), 1–13. See also Nicholas Wade, "France's All-Out Nuclear Program Takes Shape," *Science* 209 (August 22, 1980), 884–89.

[16]See Teruyan Murakami, "The Remarkable Adaptation of Japan's Economy," in Yergin and Hillenbrand, *Global Insecurity;* Richard Samuels, "The Politics of Alternative Energy Research and Development in Japan," in Ronald A. Morse, ed., *The Politics of Japan's Energy Strategy* (Berkeley: Institute of East Asian Studies, University of California, 1981), pp. 134–63; Robert Albota, "Japan's Varied Responses to Energy Vulnerability," *International Perspectives,* July–August 1981, pp. 16–19; Hirofumi Shibuta, "The Energy Crises and Japanese Response," *Resources and Energy* 5 (1983), 129–54; and Masao Sakisaka, "Energy Alternatives: Present State and Future Problems," *Journal of Japanese Trade and Industry,* November–December 1983, pp. 25–29.

stable and competitive export industries, and it was reflected in a willingness to let the economy absorb the price shocks. As John Zysman notes, "the required transfer of real resources out of consumers' pockets and into exports was accomplished quickly and with limited inflationary consequences. Overall trade surpluses were maintained and bilateral balances brought back into the black at the same time that inflation was kept below other national rates."[17] The German response relied heavily on the adjustment mechanisms of the market, with government intervening to speed the process along. Within the energy sector itself, government strategy sought to strengthen but not to control energy companies.[18]

In Japan and Germany energy adjustment was a form of industrial policy. The governments sought to work on the consumption side of the energy problem by encouraging industrial adaptation. They anticipated that a sustained and economywide adaptation would help export-oriented industries mitigate the effects of oil-price increases and that efficiency programs would reduce demand for imported oil.

EVOLUTION OF THE AMERICAN RESPONSE

The American response to the oil shocks was distinctive in several respects. First, the United States was the only country that initially sought a concerted, multilateral response to OPEC. Whereas other countries moved to adjust their internal economies to higher oil prices and to strengthen commercial ties to OPEC states, the United States worked to create a common bargaining position among consuming nations. Second, the United States was the only country that controlled domestic petroleum prices at the time of the first oil shock.[19] These price controls, the legacy of the Nixon administration's anti-inflation program, sheltered consumers from higher prices and made it difficult to use the market to push forward the process of adjustment. Under such constraints, government officials began the search for a workable response to higher oil prices.

[17]John Zysman, *Governments, Markets, and Growth: Financial Systems and the Politics of Industrial Change* (Ithaca: Cornell University Press, 1983), p. 251; see also Edward N. Krapels, *Pricing Petroleum Products: Strategies of Eleven Industrial Nations* (New York: McGraw-Hill, 1982), pp. 131–35.

[18]See Dieter Schmitt, "West German Energy Policy," in Wilfrid L. Kohl, ed., *After the Second Oil Crisis: Energy Policies in Europe, America, and Japan* (Lexington, Mass.: Lexington Books, 1982), p. 140, and U.S Department of Energy, Office of International Affairs, *Energy Industries Abroad* (Washington, D.C., September 1981).

[19]See Krapels, *Pricing Petroleum Products*.

International initiatives to coordinate the policies of industrial countries toward OPEC were launched by State Department officials in 1973, with the earliest efforts beginning prior to the October embargo. After the embargo, Secretary of State Henry Kissinger led the attempt to confront OPEC directly in order to roll back prices. Unable to achieve results through ministerial contacts with selected OPEC countries, American officials called for unity among importers. Kissinger hoped to coordinate the energy policies of the advanced industrial nations and, by so doing, jointly reduce oil consumption, thereby causing OPEC to moderate prices.

Although other Western government officials were skeptical, they met in conference at Washington in February 1974. The following November agreement was reached among seventeen advanced industrial countries (the French abstained) on the creation of the International Energy Agency. Kissinger had ambitious plans for the new international organization, but the IEA became a modest mechanism for contingency oil-sharing agreements and the exchange of information.[20] It became a focal point for the many international schemes proposed to mitigate financial and import vulnerabilities. As a device by which to recapture the erstwhile petroleum order, however, the IEA fell far short of American hopes.

American international strategy shifted. Rather than confront OPEC directly in an effort to push back prices, the U.S. government adopted a more defensive posture, trying to facilitate a dialogue between consuming and producing nations in order to establish stable international oil prices. Kissinger, hoping to protect Western investments in energy development, proposed a "common floor price" for oil. He also suggested an international consortium to pool capital for energy investment as well as to conduct joint research and development. Such agreements among the consuming nations would, he thought, permit more constructive negotiations with OPEC on prices, stable markets, and petrodollar recycling. Again his proposal met with resistance from other consuming nations. His initiative for an oil price agreement was pursued into 1975, and led to proposals to negotiate agreements on other raw materials, but proposals to negotiate

[20]Other analysts give more weight to cooperative action by industrial oil-importing nations, particularly as instituted in International Energy Agency plans and programs. See Robert O. Keohane, *After Hegemony: Cooperation and Discord in the World Political Economy* (Princeton: Princeton University Press, 1984), pp. 217–40, and Ann-Margaret Walton, "Atlantic Relations; Policy Coordination and Conflict: Atlantic Bargaining over Energy," *International Affairs* 52 (April 1976), 180–96.

stable floor prices did not provide any headway on the ultimate American goal, which was to get OPEC to moderate or reduce oil prices.

Even before these international schemes had been played out, another set of initiatives began at the national level to make the U.S. government a direct participant in domestic energy production. In effect, the government would fashion for itself a new and enlarged role in the energy sector. The initial effort, unveiled as Project Independence, was an elaborate planning program to create the statistical and informational basis for policy making on energy conservation and production. The most ambitious proposal, the Energy Resource Finance Corporation, came in 1975 from the Ford administration and involved a massive federal finance authority to direct investment into alternative energy technologies. These programs, had they been successful, would have greatly enlarged domestic energy production.

Such initiatives envisioned a major reworking of the relationship between the federal government and the energy industry, but "institutional breakthroughs" did not occur. The ability of the government to plan, coordinate, and intervene in energy markets was not significantly expanded, despite considerable efforts by government officials and politicians. By the end of the decade the U.S. government was no more weighty a presence in the energy area than it had been at the beginning.

Although long-range planning and outright control remained elusive, a third type of energy adjustment policy did take root—federal spending programs aimed primarily at pushing forward new energy technologies. By the end of the decade the government was giving massive support to energy research and development. The disbursement of funds did not challenge the private organization of energy production, but it allowed modest administrative direction over the prospective development of alternative or advanced energy technologies. Congress was comfortable with the notion of spending money, and throughout the 1970s, government expenditures on energy R & D expanded quickly.

None of these strategies addressed an important underlying problem. Price controls and regulations in place prior to the 1973 shock kept domestic oil prices lower than international prices. These price controls shielded consumers from the full brunt of higher OPEC oil costs and provided cost advantages to domestic refiners. Indirectly they subsidized consumption and discouraged domestic production. Moreover, short-term responses to market dislocations extended the price and allocation rules. As government officials confronted energy adjustment, a full-blown regulatory apparatus (and the interests it served) stood squarely in the way of policy.

A return to market pricing in petroleum was not an alternative to the other policy responses. Indeed, the State Department premised its plans for international cooperation on full adjustment of the American economy to higher international oil prices. As other strategies fell short of expectations, however, the issue of oil pricing grew in importance. Decontrol was first made a central component of adjustment strategy in 1975, but Congress vigorously opposed the Ford administration's announcement of a "presidential initiative to decontrol the price of domestic crude oil."[21] The Carter administration came to office opposing decontrol, but, to encourage conservation, quickly turned to the price mechanism. It proposed the Crude Oil Equalization Tax as a way to prevent domestic producers from reaping windfall profits while making U.S. consumers pay world prices. This imaginative attempt to create market conditions, at least on the demand side, failed before a skeptical Congress.

New international pressures in 1978 and 1979 led the Carter administration to embrace market pricing as the single most decisive form of government action available. The second round of OPEC price shocks and vigorous demands by oil-importing allies strengthened the move by executive officials toward decontrol. Oil pricing became increasingly embedded in larger and complex matters of foreign policy and international economics. In July 1978 foreign economic policy officials within the Carter administration maneuvered to link an American pledge on energy pricing to German and Japanese agreements to reflate their economies. Executive officials used foreign policy to redefine what was at stake in the decontrol proposal, effectively shifting the balance of power in the struggle among domestic groups over oil-pricing policy. These developments brought what was for energy policy the most important outcome of the decade: in April 1979 the decision to decontrol domestic prices was announced.

What had begun as an ambitious international agenda for cooperation ended with a decision to dismantle government regulation. American executive officials grudgingly scaled down their policy objectives. Unable to achieve cooperation among industrial nations, a group of executive officials had backed efforts to transform the state's role in the energy sector and to distribute massive amounts of funds

---

[21]State of the Union Message, January 13, 1975.

through existing channels. In the end, however, a government that appeared to have few alternatives embraced a return to market forces.

This final course of action had been unavailable at the time of the first oil shock, and it eventually succeeded largely because other strategies failed. Had price controls not been in place before 1973, the U.S. economy would have felt the full brunt of rising prices, which would have enforced the adjustments other nations took for granted. At the moment of crisis, however, the regulatory apparatus was too firmly established to be disassembled, and so the decontrol option was not available. The failure of other proposals in a sense cleared the ground for the market proposal; a logic of trial and error, largely implicit, was at work. The market strategy seemed to some unambitious, even a capitulation to private industry, but in fact served broad "state" goals. Moreover, the reimposition of market pricing was a formidable task for executive officials, who confronted entrenched consumer and industry interests.

The price shocks generated interlocking national and international problems. Executive officials made the national response to the energy crisis a matter of foreign policy significance, seeking to enlist allies in a coordinated Western strategy, but such initiatives hinged on the ability of public policy to influence the domestic consumption and production of energy. At various moments across the 1970s domestic and foreign policy proposals were intimately dependent on one another, even while the government departments concerned were largely separate.

What can we learn from the sequence of policy strategies? First, the shifts were not discrete, sequential movements. As noted above, international cooperative schemes envisaged substantial reductions in the American consumption of imported petroleum: an orchestrated lowering of oil imports by the industrial nations was to undercut OPEC pricing decisions. Thus the United States sought to minimize the costs of adjustment, but cooperative action would still have required government steps to dampen domestic consumption and spur production. Also, spending programs were proposed repeatedly throughout the decade, as were efforts to redefine the government's role in the energy sector, which began long before international efforts collapsed. Finally many executive officials emphasized market adjustment from the very beginning of the crisis. Decontrol as an adjustment policy had to wait, however, until domestic and international circumstances evolved to emphasize it.

Second, the shift of strategy was really a shift in emphasis and in commitment. Emphases changed as government officials ran into constraints; policy was pursued until it failed. A logic of policy is exhibited here. It is not that of straightforward rationality but more of the trial-and-error variety. Administration officials were not simply trying to develop appropriate energy adjustment policies but trying to find tools and instruments that government could effectively employ. They were asking not just what the federal government should do but also what government could do.

The American response to the oil shocks is conventionally presented as a case of policy failure, and certainly executive officials in several administrations found that their ability to implement their policy goals was severely limited. Individual proposals were blocked or significantly redesigned by congressional representatives and committees. Nonetheless, by the end of the decade the policy of oil price deregulation did signal a successful end to the search for a workable response. The question, then, is not why executive officials failed to produce an effective response but why responses unfolded as they did before terminating in a market response. Explanation requires that we explain why executive officials generated different policy proposals in a particular sequence—not just to account for a policy "outcome" but to explain both the motivation for and disposition of policy proposals. Historical cases of energy adjustment, that is to say, are part of the more general phenomenon of national adjustment to international economic change.

INTERNATIONAL CHANGE AND NATIONAL ADJUSTMENT

Many international and domestic forces set states in motion, but none is more important than the constant pressure for national adjustment to international political and economic change.[22] Constant, differential changes between national and international systems produce pressure for action. As Robert Gilpin notes, "In every international system there are continual occurrences of political, economic, and technological changes that promise gains and losses for one or another actor. . . In every system, therefore, a process of diseq-

[22]The concept of adjustment is well-established in the economics literature as part of the standard trichotomy in exchange-rate theory: adjustment, liquidity, and confidence. Benjamin Cohen traces this trichotomy back to a "celebrated international study group of 32 economists in 1964." Cohen, *Organizing the World's Money* (New York: Basic Books, 1977), p. 277 n. 18. Political scientists have also begun to use the concept of adjustment or adjustment policy. Pioneering works are Zysman, *Governments, Markets, and Growth*, esp. pp. 91–93, and Peter J. Katzenstein, *Small States in World Markets: Industrial Policy in Europe* (Ithaca: Cornell University Press, 1985), chaps. 1–2.

uilibrium and adjustment is constantly taking place."[23] This differential change may involve systemwide economic upheaval, as in the depression of the 1930s or the oil price revolution of the 1970s. It can also be more gradual, as in the changing competitive position of particular industrial sectors in advanced industrial countries. Change promises gains and threatens losses for nations and groups within nations, and for this reason states get involved in the adjustment process.[24]

Adjustment can take place in the absence of or in spite of government policy and strategy. International financial markets, for example, responded to petrodollar recycling much more effectively than Western officials anticipated, and a portion of oil price increases was absorbed in inflation. Such international market processes can reequilibrate or adjust national economies to prevailing international conditions. But even if governments direct adjustment policy primarily at the margin of larger international processes, policy is still significant. The stability and security of nations can depend on actions taken at the margin. What is of marginal importance in the long term may be of powerful significance for political and economic actors in the short term. Marginal actions stretched over extended periods, moreover, can result in profound political and economic change.[25]

We can conceive of the actions of states as controlled by strategies that states develop to cope with adjustment problems.[26] Adjustment

[23]Robert Gilpin, *War and Change in World Politics* (New York: Cambridge University Press, 1981), p. 13.

[24]See Gilpin, *War and Change;* R. J. Barry Jones, *Change and the Study of International Relations: The Evaded Dimension* (London: Pinter, 1981); Ole R. Holsti et al., eds., *Change in the International System* (Boulder, Colo.: Westview, 1980); and John A. Vasquez and Richard W. Mansbach, "The Issue Cycle: Conceptualizing Long-Term Global Political Change," *International Organization* 37 (Spring 1983).

[25]For a rather abstract discussion of system change and differential response see Herbert Simon, "The Architecture of Complexity," in *The Science of the Artificial*, 2d ed. (Cambridge: MIT Press, 1981), pp. 193–229. The issue is also discussed in terms of international equilibrium and disequilibrium. See Richard Rosecrance, *International Relations: Peace or War?* (New York: McGraw Hill, 1973), pp. 113–14.

[26]The terms "adjustment strategy" and "adjustment policy" are used interchangeably here to designate a set of policies embraced by the central executive officials of government as the chief means or organizing basis by which the government is to respond to or cope with the energy crisis. The policies themselves are proposed courses of action or inaction (e.g., laissez-faire) directed toward goals articulated by the officials. Adjustment strategy is not necessarily identical with the policy outcomes of the whole of the national government. Nor is adjustment strategy necessarily consistent with other policies of the same government. Rather, it is the articulated goals and proposed course of action of the executive officials of government. Where there are differences of opinion within the executive branch over adjustment strategy, the goals and proposals of the most senior officials and, ultimately, the president himself, are identified as government policy. See John S. Odell, *U.S. International Monetary Policy: Markets, Power, and Ideas as Sources of Change* (Princeton: Princeton University Press, 1982), pp. 15ff.

Figure 1. A typology of national adjustment strategies

	International	Domestic
Offensive	Create new international regime	Change domestic structure
Defensive	Remedial steps to defend current regime	Protect domestic structure

strategy may be directed outward at international change, or inward at domestic structural transformation, or somewhere between to maintain existing relationships.[27]

Adjustment Strategies

Solutions to adjustment problems can address the location of adjustment (international or domestic) and the objective of the solution (defensive or offensive), whether to transform the (national or international) system or to preserve the existing arrangements. This typology, summarized in Figure 1, indicates the four possibilities for states seeking to address adjustment problems.

Offensive international adjustment is the most ambitious type of response. It involves the creation of new rules of the game for international interactions; the number of states necessary to create an international regime and the required level of adherence to rules and procedures is considerably greater than for other strategies. The trade regime of the General Agreement on Tariffs and Trade, the Bretton Woods monetary system, and the proposed rules embodied in the New International Economic Order are the most ambitious examples.[28]

[27]In an analogical way Kenneth Waltz talks about internal and external means of balancing geopolitical power. Internal balancing refers to the domestic mobilization of power common to the two central states in a bipolar system. External balancing refers to the making and breaking of allies and is common to multipolar systems. The sources and implications of internal and external mobilization strategies remain an important research question. Waltz, *Theory of International Politics* (Reading, Mass.: Addison-Wesley, 1979), p. 168.

[28]It is international offensive adjustment that Stephen Krasner describes in his analysis of Third World demands for authoritative international rules and norms to cover North-South economic relations. Developing countries, he argues, are "exposed to vacillations of an international system from which they cannot extricate themselves but over which they have only limited control." Because of obdurate internal weaknesses, developing countries have consistently sought to create international regimes that will mitigate this structural vulnerability. Krasner, *Structural Conflict: The Third World against Global Liberalism* (Berkeley: University of California Press, 1985), p. 4.

The initial American response to the 1973–74 oil shock was an international offensive policy. State Department officials made attempts to forge a collective reponse by industrial oil-consuming countries, thereby inducing OPEC member governments to moderate or roll back oil prices. Such was the rationale of American executive officials: collaborative efforts to restrain oil imports and to foster alternative sources of energy, they argued, would give the industrial countries a stronger position from which to bargain with OPEC over prices. The strategy was directed at reestablishing the old petroleum order. In effect, the initial U.S. response to unanticipated international change was to go to the source of that change and reverse its course.

If offensive international adjustment seeks to alter the sources of change, defensive international adjustment involves the use of international agreements to moderate international economic change, in effect forging international agreements to protect existing domestic industries and institutions. The Multi-Fiber Agreement, for example, has evolved over several decades into an elaborate system whereby textile-producing nations distribute market shares. Change in the sector stems from shifting comparative production costs and technological advantages among and between advanced and developing nations. The agreement, initially sought by the United States in an effort to protect domestic textile producers, mandates market shares in a wide range of textile goods. In recent years a multitude of commercial agreements has been negotiated in various trade sectors to accomplish the same ends. These agreements frequently take the form of negotiated trade quotas and orderly marketing agreements.[29]

With the failure to develop a collaborative strategy to confront OPEC directly, State Department officials moved to an international defensive strategy. The United States now aimed to facilitate a di-

[29]On the MFA see Vinod K. Aggarwal, *Liberal Protectionism: The International Politics of Organized Textile Trade* (Berkeley: University of California Press, 1985); more generally, David Yoffie, *Power and Protectionism: Strategies of the Newly Industrializing Countries* (New York: Columbia University Press, 1983), and also Susan Strange and Roger Tooze, eds., *The International Politics of Surplus Capacity* (London: Butterworth, 1980). In some respects the difference between international offensive and international defensive strategies is captured in Robert Keohane's distinction between "control-oriented" regimes and "insurance" regimes. Keohane, "The Demand for International Regimes," *International Organization* 36 (Spring 1982), 325–55. Control-oriented regimes seek to prescribe and regularize behavior among participants within the regime. Members gain a high measure of control over the behavior of other actors. Insurance regimes are less ambitious and seek to pool risks of specific international economic change among regime participants. Whereas control-oriented regimes attempt to go to the source of deleterious international change, insurance arrangements aim to soften the impact of that change.

alogue between consuming and producing nations in order to stabilize international prices. To protect massive Western investments in energy development, Secretary of State Kissinger proposed a common price floor for oil, which would strengthen the reliability of petroleum prices and supplies over an extended period of time. The proposal was defensive in that the United States was now willing to countenance higher oil prices. The agreement sought to moderate price increases and gain more time for industrial importing nations to make domestic adjustments.

Domestic offensive strategy changes the structure of national industries and institutions in an effort to cope with international economic change. It may involve phasing out or encouraging the growth of particular industries or creating new arrangements that facilitate domestic economic adjustment. Governments focus their efforts inward, adapting institutions and redeploying resources. One form of this strategy may involve vigorous and anticipatory government action to gain an edge over competitors. Investment in R&D, rationalization of corporate decision making on investment, and other government tactics encourage or coerce private firms to alter their behavior. In another form of the strategy, the government may stand aside and let market forces transform the economy. Here the state acts by not acting; the state, a "gatekeeper," chooses to keep the gate open.[30]

In responding to higher oil prices, the European and Japanese governments primarily pursued domestic offensive policies. These efforts took many forms, ranging from passive measures that allowed international oil prices to move through the domestic economy to aggressive programs of industrial restructuring. The Japanese government, for example, phased out such energy-intensive industries as aluminum and petrochemical production. The West German government was less interventionist, acting primarily to ensure that industrial producers and energy consumers weighed the full costs of increased energy prices. Similarly, the American decontrol of oil prices was a domestic offensive policy, though it emerged much later than did policies in Europe and Japan.

Domestic defensive adjustment is a protective strategy that seeks to avoid change altogether, and it typically culminates in the erection of

---

[30]See the discussion of Japanese "anticipatory adjustment policy" in Michèle Schmiegelow, "Cutting across Doctrines: Positive Adjustment in Japan," *International Organization* 39 (Spring 1985), 261–96. The OECD uses several categories of adjustment to characterize national patterns in the 1970s and beyond. See OECD, *Positive Adjustment Policy—Managing Structural Adjustment* (Paris, 1983).

barriers to new forms of change. The tariff is the most obvious state action, but subsidies and other nontariff barriers conform to the same defensive strategy.[31] Throughout much of the 1970s the American maintenance of price controls was a domestic defensive response to international charge.

CONCLUSION

Two crucial questions emerge from this sketch of energy adjustment in the 1970s. First, why did industrial countries pursue divergent energy adjustment policies, and what were the implications of these divergent policies for international cooperation and conflict? We need to know why the United States pursued a collaborative policy in the initial stages of the crisis and why other industrial countries pursued independent national policies.

Second, why did American energy adjustment evolve through a *sequence* of policy initiatives, from international to national, and from interventionist to market-based? We need to explain why American executive officials initially sought international cooperation. Once energy adjustment policy had moved to the domestic level, moreover, we need to explain the sequence of policies observed.

This search for explanations for the style and evolution of adjustment policies, in their international and domestic guises, centers on the "weightiness" or "capacity" of the state. The key is the Machiavellian question concerning the constraints and opportunities for states to realize the goals they embrace. The notion of state capacity, according to Theda Skocpol, refers to the ability of states "to implement official goals, especially over the actual or potential opposition of powerful social groups or in the face of recalcitrant socioeconomic circumstances."[32] At moments of crisis, what can states accomplish? What international and national resources—power—can state officials marshal in the face of international dilemmas? How do domestic structures shape the ability of states to achieve official objectives?

[31]Tariffs need not always be associated with defensive domestic adjustment. When used to nurture infant industries, tariffs may help transform the industrial structure of a nation. Similarly, investment in research and development can be defensive rather than offensive in its design.

[32]Theda Skocpol, "Bringing the State Back In," in Peter Evans, Dietrich Rueschemeyer, and Skocpol, eds., *Bringing the State Back In* (New York: Cambridge University Press, 1985), p. 9. See also the concluding essay in that volume and Eric Nordlinger, *On the Autonomy of the Democratic State* (Cambridge: Harvard University Press, 1981), pp. 8–28.

These questions about state capacity require attention to the interaction between state officials and their environment.

The American response to the oil crises provides a window on to the more general attempts of the United States to cope with transformations in the postwar international system. The problems American government officials encountered in asserting international leadership in the midst of the oil crises forced officials to make difficult internal adaptations. As the international political and economic position of the United States continues to decline, these difficult internal adjustments are becoming the grist of domestic politics and a decisive constraint on foreign economic policy. The oil shocks forced the U.S. government to explore the scope and limits of its international and domestic capabilities.

CHAPTER TWO

Explaining Energy Adjustment Policy

> The difference between international politics as it actually is and a rational theory derived from it is like the difference between a photograph and a painted portrait. The photograph shows everything that can be seen by the naked eye; the painted portrait does not show everything that can be seen by the eye, but it shows, or at least seeks to show, one thing that the naked eye cannot see: the human essence of the person portrayed.
>
> Hans J. Morgenthau, 1960

Disruptions in international oil markets and rapidly escalating prices during the 1970s generated a wide range of socioeconomic challenges to which government leaders in the industrial world were forced to respond. The unfolding events of that decade prompt two questions. First, why did the United States campaign for a concerted international response while officials in the other major nations were emphasizing narrower, national programs? Differences in policy reflected both distinctive conceptions of what was at stake and divergent capacities for solving energy and related problems. They also had fateful implications for the inclination of states to join in cooperative actions that would foster mutual gains. Second, why did American officials move from international to domestic and from interventionist to market policies? A process of trial and error was at work. What motivated executive officials to move through a sequence of proposals, and why did some ideas fail and others succeed?

These questions, and the larger issue of the capacity of states to cope with international political and economic change, require a theoretical appreciation of the bases of state action. Various literatures address the explanation of policy outcomes. Building upon several promising lines of analysis, I present an institutional approach to understanding adjustment policy.

APPROACHES TO FOREIGN ECONOMIC POLICY

In seeking to explain the actions of states, analysts make a basic distinction between sources of state behavior that emanate from the international system and those that come from the domestic system.[1] "Systemic" explanations trace state actions to the structure of incentives and constraints created by the international system as a whole. Societal explanations trace state actions to pressures and constraints generated within the domestic political system, where theorists have attempted to explain foreign economic policy by focusing on the play of private group or class interests ("society-centered" explanations) or by focusing on the shaping and constraining influence of government institutions and the officials within them ("state-centered" explanations).

Systemic Explanations

The international system, according to systemic theorists, generates enduring and powerful pressures on nation-states and, consequently, on foreign economic policy. The basic features of the international system of concern to these theorists are found within three arguments about the international system central to the Realist tradition of analysis. First, the international system is dominated by sovereign states, each beholden to no higher authority than itself nor to any purpose higher than the protection of the nation's territorial integrity and material well-being. Second, the primacy of states and their substantive interests stem from the anarchy of the international system. Cooperative agreements and mutual restraint may flourish at various historical moments, but behind international cooperation is the enduring absence of authority to enforce agreement. Finally, in a system so constituted, states tend to behave purposively and pursue what their representatives perceive to be national or state interests.[2]

[1] See Kenneth Waltz, *Theory of International Politics* (Reading, Mass.: Addison-Wesley, 1979).
[2] See Robert O. Keohane, "Theory of World Politics: Structural Realism and Beyond," in Ada W. Finifter, ed., *Political Science: The State of the Discipline* (Washington, D.C.: American Political Science Association, 1983); John A. Vasquez, *The Power of*

The structure of the international system has a basic and relentless impact on state action. It should be understood, according to Kenneth Waltz, in terms of an ordering principle (e.g., anarchy) and a particular distribution of power. The international system is similar in structure to the market in that both are systems created by self-regarding actors. The system, Waltz argues, is the unintended yet inevitable and spontaneously generated outgrowth of activities by nation-states concerned fundamentally with their own survival. Different types of systems generate different international outcomes; multipolar systems, for example, tend to be less stable than bipolar ones. Recurring patterns of international behavior are explained in terms of the enduring organizational structure of the international system.

Other systemic theorists address more proximate international outcomes, such as policy coordination and the creation and maintenance of regimes. The theory of hegemonic stability, for example, attempts to account for such outcomes as international economic openness and regime strength in terms of the distribution of economic capabilities among the major powers of the system. Because of the power and interests of dominant states, it is hypothesized, hegemony will lead to openness and stable regimes.[3] Some systemic theorists have gone beyond general incentives and attempt to specify more precise sets of state goals. In developing a theory of "international economic structure," for example, David A. Lake constructs a deductive framework that specifies variations in expected national trade strategies based on the relative international economic position of nations. The relative size and productivity of national economies provide the crucial situational variables that weigh on the likely trade strategy chosen by particular nations.[4]

Changes in the behavior of nation-states, and in system outcomes, are explained by systemic theorists in terms of changes not in the internal characteristics of nation-states but in the system itself or in

Power Politics (New Brunswick: Rutgers University Press, 1983); and Robert G. Gilpin, "The Richness of the Tradition of Political Realism," *International Organization* 38 (Spring 1984), 287–304.

[3] See Robert O. Keohane, "The Theory of Hegemonic Stability and Changes in International Economic Regimes, 1967–1977," in Ole Holsti et al., *Change in the International System* (Boulder, Colo.: Westview, 1980), pp. 131–62; Stephen Krasner, "State Power and the Structure of International Trade," *World Politics* 28 (April 1976), 317–43; and Krasner, ed., *International Regimes* (Ithaca: Cornell University Press, 1983).

[4] David Lake, "Beneath the Commerce of Nations: A Theory of International Economic Structures," *International Studies Quarterly* 28 (1984), 143–70; "International Economic Structures and American Foreign Economic Policy, 1887–1934," *World Politics* 35 (July 1985), 517–43; and *Power, Protection, and Free Trade: International Sources of U.S. Commercial Strategy, 1887–1939* (Ithaca: Cornell University Press, 1988).

the relative position of particular nation-states. Waltz argues that "it is not possible to understand world politics simply by looking inside of states," but systemic theory is constructed at such a level that it does not require the analyst to look inside the nation-state at all. This level of analysis provides the situational context for state action, by setting the outer boundaries within which policy must be constructed and sustained.[5]

Systemic theory does not intend to explain foreign policy. The international system generates incentives and constraints; it rewards and punishes the actions of states.[6] But it does not provide a theory of state action as such. More differentiated systemic theories, such as Lake's, do provide specific sets of national preferences that self-interested states are likely to follow. Once again, however, the ability of states to perceive and act upon those interests hinges on domestic circumstances. The virtue of the international-centered approach is that it provides the situational context for foreign economic policy. Explanation of actual policy outcomes requires additional conceptual variables that reside within nation-states.

Society-Centered Explanations

Societal explanations trace outcomes back to the domestic forces or political groups that lay claim to foreign economic policy. Societal interests and groups, and their impact on policy outcomes, are conceptualized in various ways. Two prominent society-centered approaches stress different elements within society that influence or shape policy outcomes.

Drawing on pluralist theory, the interest-group approach focuses on the play of organized groups within the policy process. Policy is the outcome of the competitive struggle among groups for influence over specific policy decisions. Government institutions provide an arena for the competition among groups and do not decisively bias the decisions that emerge. In its simplest formulation, this approach anticipates a spontaneity and fluidity in the involvement of interest

[5]Waltz, *Theory*, p. 65. See the critique of systemic theory in Robert O. Keohane, *After Hegemony* (Princeton: Princeton University Press, 1985), pp. 25–29.

[6]The impact of international structure on state action is described in this literature in terms of constraints. Waltz argues that "actors may perceive the structure that constrains them and understand how it serves to reward some kinds of behavior and to penalize others. But then again they either may not see it or, seeing it, may for any of many reasons fail to conform their actions to the patterns that are most often rewarded and least often punished. . . . The game one has to win is defined by the structure that determines the kind of player that is likely to prosper." Waltz, *Theory*, p. 92.

groups in policy making. Various types of groups—industry associations, unions, consumer groups—are activated and form coalitions on the basis of the particular issues at stake. As issues change, so do the groups involved and their coalitions. This pluralist formulation does not directly relate interests and the involvement of groups in policy making to broader social structures. Explanations for policy outcomes emerge from the societal interests at stake and the groups that effectively organize to get involved in decision making.[7]

This approach has been used widely in the analysis of foreign economic policy. Various scholars have explained American trade policy, for example, in terms of the domination of private groups.[8] Government is understood to be primarily passive, either providing a neutral arena or acting as a disinterested referee. Societal groups make demands, government supplies policy; policy is the outcome of shifting group coalitions. If the character of societal interests is understood, the explanation of policy follows easily.

A second approach, developed to explain policy outcomes in comparative perspective, employs a larger-scale and more structured conception of societal interests and groupings. Most prominently, Peter Gourevitch has traced distinctive national economic policies during the Depression of the 1930s to systemic differences in the position of (broadly drawn) social and industrial groups. The major social groups he identifies—agriculture, industry, and labor—are further divided in terms of international competitive position, and resulting model of societal or sectoral interests provides the basis to explore cross-national variations in coalitional limits and possibilities. Shifts in policy are traced back to shifts in sectoral preferences (understood as a function of the sector's changing "situation" within the international economy) and the emergence of cross-sectoral coalitions. When societal interests and groupings are understood, an explanation of policy follows.[9]

[7]The seminal statements of the pluralist perspective are Arthur F. Bentley, *The Process of Government* (Chicago: University of Chicago Press, 1980); David B. Truman, *The Governmental Process: Political Interests and Public Opinion* (New York: Knopf, 1951); and Robert A. Dahl, *Who Governs?* (New Haven: Yale University Press, 1963). Subsequent revisions of the pluralist model of policy making dispute many of its elements but retain the basic society-centered focus. See Theodore J. Lowi, *The End of Liberalism* (New York: Norton, 1969), and Charles E. Lindblom, *Politics and Markets* (New York: Basic Books, 1977).

[8]See, for example, E. E. Schattschneider, *Politics, Pressures, and the Tariff* (New York: Prentice-Hall, 1935); Raymond A. Bauer, Ithiel de Sola Pool, and Lewis Anthony Dexter, *American Business and Public Policy: The Politics of Foreign Trade*, 2d ed. (Chicago: Aldine-Atherton, 1972); and Peter F. Cowhey and Gary C. Jacobson, "The Political Organization of Domestic Markets and U.S. Foreign Economic Policy," University of California, San Diego, unpublished paper, 1984.

Society-centered explanations locate the domestic groups and coalitions that gain advantage from particular policy outcomes and that may support or sustain those policies. Yet they do not help us when the interests and capacities of groups are themselves influenced or shaped by larger domestic institutional structures. The organizational structure of the state, and the preference of administrators and politicians who occupy positions within that structure, may weigh heavily in policy outcomes. State structures, by shaping the institutional terrain where policy struggles are played out, can have an important if indirect influence on access to decision making and the resources that bear on decision making. State officials also may blunt, shape, or ignore the activities of social groups. An approach that focuses exclusively on social groups, in sum, captures only the demand for policy, not its supply.

To understand how policy responds to societal demands, we need a more explicit understanding of the "black box" of the state. In trade policy, for example, the responsibility for decision making was gradually transferred from the Congress to the executive branch in the decades following the disastrous Smoot-Hawley tariff of 1930. Interest-group demands for protection did not abate, but executive institutions were less responsive to those demands.[10]

State-Centered Explanations

A substantial literature incorporates the state into societal understandings of foreign economic policy, developing state-centered explanations in two ways. First, unlike the interest-group approach, it finds the state able to develop and implement autonomous preferences, even in the face of pressure from private interests. Foreign economic policy is not simply the hostage of societal groups; the state

[9]Peter Gourevitch, *Politics in Hard Times: Comparative Responses to International Economic Crises* (Ithaca: Cornell University Press, 1986). For a similar approach see Tom Ferguson, "From Normalcy to New Deal: Industrial Structure, Party Competition, and American Public Policy in the Great Depression," *International Organization* 38 (Winter 1984), pp. 41–94. This societal approach is compatible with an instrumental neo-Marxist orientation. See Jeff Frieden, "From Economic Nationalism to Hegemony: Social Forces and the Emergence of Modern U.S. Foreign Economic Policy, 1914–1940," *International Organization* 42 (Winter 1988).

[10]See Robert Pastor, *Congress and the Politics of U.S. Foreign Economic Policy, 1929–1976* (Berkeley: University of California Press, 1980); Judith Goldstein, "Ideas, Institutions, and Trade Policy," and Stephan Haggard, "The Institutional Foundations of 'Hegemony': Explaining the Reciprocal Trade Agreements Act of 1934," both in *International Organization* 42 (Winter 1988); and I. M. Destler, *American Trade Politics: System under Stress* (Washington, D.C.: Institute for International Economics, 1986).

may be able to resist or blunt interest-group or class demands in favor of its own independent set of policies. Attention focuses on state officials who, pursuing their own agenda, develop and seek to implement an autonomous set of preferences. Second, the organizational structure of the state may have broader and often unintended effects on the interests and capacities of individuals and groups—private groups and public officials alike—and, consequently, on policy outcomes. The state, in this view, is a set of well-entrenched institutions that undergird the policy process and influence the outcomes.

The first line of inquiry focuses on the autonomy of goals embraced by state officials and their impact on policy. In American policy on raw materials, for example, Stephen Krasner identified a set of peculiarly "state" goals that systematically triumphed over the competing interests of private corporations.[11] The autonomous actions of the state came from decision makers in the executive who, by virtue of their insulated position, were able to conduct foreign policy in accordance with a stable set of goals. Government institutions are not simply an arena of policy conflict from this perspective; they also provide platforms from which executive officials can pursue distinctive goals that are not easily traceable to the interests or activities of private groups and may in fact provoke powerful societal opposition.

The autonomy of state officials may also stem from the role that policy experts and bureaucrats play in the policy process. In the emergence of welfare policy in Britain and Sweden, for example, Hugh Heclo argues, civil servants were instrumental in developing and implementing breakthroughs in social policy. Many forces influenced policy—stages of economic development, struggles of interest groups and political parties, leadership of government administrators—but Heclo concludes that the bureaucracies in both countries were the leading agents of policy innovation.[12]

If state officials have some degree of autonomy, what are the origins of state interests?[13] One response stresses the importance of bureaucratic or narrow organizational interests. The maintenance or

[11]Stephen D. Krasner, *Defending the National Interest* (Princeton: Princeton University Press, 1978).

[12]Hugh Heclo, *Social Policy in Britain and Sweden* (New Haven: Yale University Press, 1974), p. 301.

[13]Peter Gourevitch, for example, argues that "the basic problem with this line of reasoning is that it provides no explanation for the orientation of state policy in the supposedly state-dominated countries." See "The Second Image Reversed: The International Sources of Domestic Politics," *International Organization* 32 (Autumn 1978), 903; see also Philippe Schmitter, "Neo-corporatism and the State," in Wyn Grant, ed., *The Political Economy of Corporatism* (New York: St. Martin's, 1985).

expansion of bureaucratic missions and control over budgetary resources, according to this perspective, lead to systematic and autonomous government actions.[14] Another response suggests the importance of bureaucratic problem solving and social learning. Government officials, Heclo argues, are as much engaged with "puzzling" as they are with "powering." Policy innovations may emerge as straightforward efforts by officials to anticipate or react to a variety of socioeconomic crises and dilemmas.[15] A third response focuses on the unique foreign policy mandate of executive officials.[16] Organizational position within government is decisive; it is why these officials tend to hold policy views that differ from those of other officials within government and from those of societal groups. None of these three approaches gives clear-cut, a priori predictions as to when and in what ways government officials will pursue autonomous goals. In each case the question remains empirical, a matter of reconstructing the actual behavior of government officials.

The broader organizational structures of the state, within which bureaucratic and executive officials operate, form the focus of a second state-centered literature. It is concerned with the origins and dynamics of state structures and with the impact of these organizational structures on policy outcomes. This state-centered literature is rooted in Max Weber's political sociology of the state. Weber understood the modern state to be a compulsory association claiming sovereign and coercive control over a specific territory and population.[17] It is distinguished from other social organizations by its monopoly claim on the legitimate use of violence. Force is not the only means available to the state, and in most countries it is rarely exercised, but it is the only means specific to the state. All other instrumentalities are built around it. Within a defined territory, the state's monopoly on force is the precondition for the sanctioning of law and property rights and, ultimately, for resolving political conflict. Everything else follows from this circumstance.[18]

Weber grants the state a significance unto itself as a set of relatively

[14]See, e.g., Graham Allison, *Essence of Decision: Explaining the Cuban Missile Crisis* (Boston: Little, Brown, 1971), and Morton H. Halperin, *Bureaucratic Politics and Foreign Policy* (Washington, D.C.: Brookings, 1974).

[15]Heclo, *Modern Social Policy*, p. 305.

[16]See especially Krasner, *Defending the National Interest.*

[17]See Max Weber, *Economy and Society,* ed. Guenther Roth and Claus Wittich (Berkeley: University of California Press, 1978), p. 56.

[18]Max Weber, "Politics as a Vocation," in *From Max Weber: Essays in Sociology* (New York: Oxford University Press, 1946), p. 78. For an elaboration of this view see Peter P. Nicholson, "Politics and Force," in Adrian Leftwich, ed., *What Is Politics? The Activity and Its Study* (Oxford: Blackwell, 1984), pp. 33–45. The point is captured by Samuel E. Finer: "Tell a man today to go build a state; and he will try to establish a definite and

differentiated organizations that are not reducible to or merely reflect socioeconomic setting. Built around legal and bureaucratic institutions, the state both is distinct from and interacts with other economic and social structures.[19] The independence of the modern state from society and economic classes is bound up with the search for legitimate forms of political domination and the resulting emergence of legal-rational authority. Comprehensive legal rules, manifest in formal government organizations, provide legitimacy to the stable set of expectations rulers and ruled alike attach to the exercise of political power.[20]

The Weberian perspective leads to the identification of the general characteristics of states, but recent theorists explore variations in the institutional characteristics of states and their interrelationships with larger political, economic, and social structures. They are concerned with specifying the nature of and variations in state capacities. The goals state officials develop and their ability to carry them out depend on existing institutional tools and resources.[21] In addition, the interests and capacities of social groups and classes are shaped and influenced by prevailing organizational structures of the state.[22]

Fundamental to the state-centered perspective is the belief that government activities and public policy are not simply expressions of

---

defensible territorial boundary and compel those who live inside it to obey him. Having done this he will have founded his *State*." Finer, "State-building, State Boundaries and Border Control," *Social Science Information* 13 (1974), 79.

[19]The Weberian distinction between state and society is noted by Randall Collins: "Individuals or groups are coordinated on two analytically distinct principles, which correspond to the spheres of 'society' and 'state.' In the sphere of 'society,' groups are formed as 'constellations of interests,' in which the parties act together voluntarily for what they feel is their mutual benefit. Such groups are formed on two bases: coincidence of interests in the economic market, and feelings of identity with others who hold a common culture or ideal. In the sphere of the 'state,' coordination is based on domination, in which one individual or group is placed in a position to enforce his will on the others. Such coordination is based on an organizational apparatus of domination and on some kind of principles of legitimacy. In both spheres of action, there is a struggle for advantage. . . . Individuals struggle for advantage within organizations, and organizations struggle with each other." Collins, "A Comparative Approach to Political Sociology," in Reinhard Bendix, ed., *State and Society: A Reader in Comparative Political Sociology* (Berkeley: University of California Press, 1968), pp. 48–49.

[20]Weber, *Economy and Society*, 1:212–26.

[21]See Peter J. Katzenstein, ed., *Between Power and Plenty: Foreign Economic Policies of Advanced Industrial States* (Madison: University of Wisconsin Press, 1978), and John Zysman, *Governments, Markets, and Growth* (Ithaca: Cornell University Press, 1983).

[22]E.g., Theda Skocpol and Kenneth Finegold, "State Capacity and Economic Intervention in the Early New Deal," *Political Science Quarterly* 97 (1982), 255–78; Margaret Weir and Skocpol, "State Structures and the Possibilities for 'Keynesian' Responses to the Great Depression in Sweden, Britain, and the United States," in Peter Evans, Dietrich Rueschemeyer, and Skocpol, eds., *Bringing the State Back In* (New York: Cambridge University Press, 1985), pp. 107–63.

societal demands or straightforward responses to international political and economic conditions. The organizational structures of the state and the administrators and politicians within them exert powerful influences on the circumstances and activities of policy making. Organizational structures influence the types of policies likely to be generated and successfully implemented by influencing the access groups and individuals have to policy making and the resources they are able to wield. State structures thus affect the capacity of various individuals and groups to pursue their goals. Moreover, politicians and executive officials are positioned to pursue policy goals that do not merely reflect social interests. These state officials, enabled and constrained by organizational structures, may blunt, reshape, or ignore social demands on policy. At the same time, they are well placed to interpret and act upon the pressures and opportunities that arise from the nation's changing international position.

The strength of this structural, state-centered approach is its focus on enduring institutions within which government officials and private groups and individuals are situated. For those interested in explaining policy outcomes, however, the approach has weaknesses as well. An exclusive focus on state structures may obscure important interactions between public and private actors. The organizational characteristics of societal groups and classes may be as important as the formal characteristics of the state to the ability of state officials to implement policy. Moreover, the structural orientation of the approach may obscure the importance of individual behavior and of political process in shaping specific outcomes.[23] The organizational structures of the state may constrain individuals and groups, but they do not exclude individual action. At particular historical junctures those organizational structures may in fact be altered or transformed.

I find the state-centered line of inquiry promising because of the questions it asks. A focus on foreign economic policy (or, more specifically, energy adjustment) leads to questions about state capacity, about the ability of government officials to develop and implement effective policy in order to deal with large-scale international political and economic problems. State capacity must, however, be formulated in such a way as to allow societal and international variables to be integrated into explanations of policy outcomes.

---

[23]This argument is made by Peter Hall, *Governing the Economy: The Politics of State Intervention in Britain and France* (New York: Oxford University Press, 1986). For a critique of "state-centered structuralism" see Margaret Levi, *Of Rule and Revenue* (Berkeley: University of California Press, 1988).

AN INSTITUTIONAL APPROACH TO POLICY EXPLANATION

The perspective I am developing is an institutional approach. As I argued earlier, the most promising analysis does not simply replace societal and systemic explanations with state-centered explanations. Rather, the task is to appreciate the interaction of these variables and the manner in which the organizational features of the state and the activities of executive officials and politicians mediate larger societal and international forces. This approach is "institutional" because it conceives of mediation and interaction as grounded in institutional relationships that persist over time and that are relatively resilient against the idiosyncratic actions of groups and individuals.[24]

I follow three lines of this institutional approach as they relate to the politics of energy adjustment. First, I focus on the manner in which political and economic crises reveal the forces that sustain and reshape the organizational structures of the state. Second, I examine the types of state actions that are likely to follow from variations in the larger domestic and international setting of the state. Finally, I examine variations in the capacities of state officials and in their organizational underpinnings.

State Formation and U.S. Intervention in the Energy Sector

All states have core organizations that carry out fiscal, coercive, judicial, and administrative activities within their domain. Variations abound, however, in the activities and organizational structures of states, both between states and within states over time. The role of representative or parliamentary bodies, the centralization and coherence of bureaucratic organizations, and the political resources and access to decision making of executive officials are all aspects of the prevailing organizational structures of the state. At this general level, therefore, we are looking at the "architecture" of the state and its interlocking parts.

State structures are important because they tend to persist over long periods of time and because they serve to shape and constrain the actions of groups and individuals struggling over public policy.

---

[24]See the discussion of this approach in James March and Johan P. Olsen, "The New Institutionalism: Organizational Factors in Political Life," *American Political Science Review* 78 (September 1984), 735; Stephen D. Krasner, "Approaches to the State: Alternative Conceptions and Historical Dynamics," *Comparative Politics* 16 (January 1984), 223–45. Peter Hall also describes his enterprise as an institutional approach, in *Governing the Economy*.

Politicians and bureaucrats at any given moment are not in a position to create state capacities to meet specific challenges. The institutions within which these individuals operate and the legacies of previous policy constrain the possibilities for current policy action. Because institutions are neither fluid nor able to respond to the immediate needs of state officials, they exert a powerful influence on what officials can actually accomplish.[25]

A variety of historical forces has shaped the origins and trajectories of modern states and given them distinctive organizational characteristics. Variations in the sequence and timing of political and economic development and the state-building responses to economic depressions and wars have had powerful effects on representative and bureaucratic organizations. In advance of the spread of democratic institutions, for example, many European nations constructed powerful administrative organizations that strengthened the role executive officials could play in subsequent periods of economic and political development.[26] War and geopolitical conflicts in the nineteenth and twentieth centuries also helped centralize European state bureaucracies, creating incentives for the development of extensive capacities for economic intervention and extraction.

In the United States the constitutional mandate of the nation's founding preserved a loose federal system that dispersed sovereignty across national, state, and local levels and among judicial, congressional, and executive branches. In the nineteenth century, when European state bureaucracies were expanding, the United States remained a government of courts and parties. Highly competitive political parties and patronage-oriented democratic politics strengthened the role of congressional-centered government. The spread of a mass-based, democratic political system in the United States preceded the establishment of centralized administrative institutions, unlike in Europe, and this sequence of political development served to constrain bureaucratic centralization throughout the nineteenth century.[27]

[25]The inertial characteristics of institutions are discussed in the Conclusion. See also G. John Ikenberry, "Conclusion: An Institutional Approach to American Foreign Economic Policy," *International Organization* 42 (Winter 1988).

[26]See Charles Tilly, ed., *The Formation of National States in Western Europe* (Princeton: Princeton University Press, 1975); Gianfranco Poggi, *The Development of the Modern State* (Stanford: Stanford University Press, 1978); Reinhard Bendix, *Kings and People: Power and the Mandate to Rule* (Berkeley: University of California Press, 1978); and Michael Mann, "The Autonomous Power of the State: Its Origins, Mechanisms and Results," *Archives européennes de sociologie* 24 (1984), 185–213.

[27]See Stephen Skowronek, *Building a New American State: The Expansion of National Administrative Capacities, 1877–1920* (New York: Cambridge University Press, 1982),

The relatively isolated economic and geopolitical position of the United States in the decades prior to World War I also reinforced a congressionally dominated political system. Large internal markets and the absence of a large, capable government bureaucracy permitted hierarchical and integrated business enterprises to grow in the late nineteenth century. Whereas European nations developed extensive institutional relationships between business and the state in this formative industrial period, American public officialdom played a subsidiary role in economic development.[28]

The basic organizational features of the American state were built upon but largely unchanged during the world wars and Depression of the twentieth century. Involvement in the world wars did not have the same effect on state building in the United States as it did in Continental Europe. America mobilized for war with a variety of temporary administrative programs that brought private business directly into the offices of the state. These state-sponsored, but privately run, emergency programs were easily disassembled in the postwar years, largely preserving the decentralized organizational features of the state.[29]

This preservation of the state's decentralized and fragmented organizational structure is striking in comparative perspective, but some islands of bureaucratic capacity did emerge. In the 1930s the Depression and a changing international economic environment provided an impetus for executive centralization in trade policy. Passage of the Reciprocal Trade Agreements Act of 1934 (RTAA) signaled a long-term transfer of responsibility for trade policy from Congress to the executive. The result was a decline in the direct influence that societal groups could wield over trade policy. Later, World War II put the United States in a position to act upon the new set of international demands and opportunities, and the crisis of war and the subsequent emergence of Soviet-American hostilities had catalytic effects on foreign and economic policy institutions and the power of the executive. Franz Schurmann argues that there were systematic interactions between war, ideology, and executive power. The crisis of war acted to

---

and J. Rogers Hollingsworth, "The United States," in Raymond Grew, ed., *Crises of Political Development in Europe and the United States* (Princeton: Princeton University Press, 1978).

[28]See David Vogel, "Why Businessmen Distrust Their State: The Political Consciousness of American Corporate Executives," *British Journal of Political Science* 8 (1978), 45–78. See also Andrew Shonfield, *Modern Capitalism: The Changing Balance of Public and Private Power* (New York: Oxford University Press, 1965), pp. 298–329.

[29]For a discussion of wartime mobilization, see Robert A. Dahl and Charles E. Lindblom, *Politics, Economics, and Welfare* (New York: Harper, 1953), pp. 402–12.

"increase the power of the executive levels of organization." At the same time the unrivaled international position of the United States in the later stages of the war and its aftermath gave Presidents Franklin Roosevelt and Harry Truman extraordinary opportunities to articulate a broad "ideological vision" that also strengthened the role of the executive.[30]

Conceptions of organizational structure are useful in situating the role and capacities of groups and state officials as they struggle over policy at particular historical periods. Left at this architectural level, however, they assume that groups and individuals are automatically channeled into courses of action that conform to the overarching organizational structure of state and society. We need also to focus on moments when groups and individuals seek to change or overcome institutional constraints, and in particular on efforts by executive officials and other individuals to build new state capacities. An understanding of the dominance of a private system of petroleum production and distribution in the United States and the "weak" position of executive officials must take into account the timing and phases of industrial and bureaucratic development in the decades spanning the turn of the century. Several critical periods of American political and economic history shaped for decades the structure of relations between the private petroleum industry and public officials.

State involvement in the petroleum industry was fixed in organizational structures that were not easily altered. Government-industry relations in the industry conformed to a more general pattern in which the state's regulatory involvement in industrial activities was highly circumscribed. This pattern, striking in comparative perspective, is part of the broad historical development of business and state organization in the decades spanning the turn of this century. Distinctive are the rise of big business organization, conflict between large and small firms, and the limited techniques available to the state in its efforts to get involved in industrial affairs. Government-business relations were also fixed in ways specific to the petroleum industry, because of the international and domestic evolution of the industry, the state's involvement in mediating industrial conflicts, and the periodic concern of state officials with issues of energy supply and energy security.

Attempts to extend state capacities in the energy sector were sporadic and fragmentary, an exercise that left executive officials with

[30]Franz Schurmann, *The Logic of World Order* (New York: Pantheon, 1974), pp. 8, 21–22.

34

few planning capacities and mechanisms for involvement. At moments of crisis for national security, executive officials made efforts either to become more directly involved in energy production or to develop added planning capacity. Such initiatives were generally blocked or dismantled after the crisis had passed. Institution building did not survive the crises that provoked it.

The Janus-Faced State and Strategies of Adjustment

States participate in both domestic and international political systems. Made up of different organizational capacities and modes of operation, states occupy a unique position to mediate internal and external change—J. P. Nettl's image of a gatekeeper is apt.[31] The international and domestic activities and capacities of states are inextricably linked. Untangling the logic of the state's Janus-faced relationship with its international and domestic settings provides an initial understanding of the bases of state action.

Although states vary widely, all are composed of organizations that make exclusive claims to the sovereign and coercive control of a territory and its population. These monopoly claims legitimate a variety of core activities: the raising and disbursement of revenue, the promulgation and enforcement of rules and property rights, and the maintenance of political order. States also claim to define and represent broad national interests and seek to protect and advance those interests in the face of international competition.[32]

States are thus separate from but enmeshed in a complex set of relationships with society. It is a conception of the state captured most effectively in historical analyses of states at their moments of birth. Charles Tilly has sketched a series of relationships between war making, capital accumulation, extraction, and European state making. He notes that leaders of nascent states, engaged in war with adjacent powerholders, needed to extract resources from local producers and traders: "The quest inevitably involved them in establishing regular access to capitalists who could supply and arrange credit, and to imposing one form of regular taxation or another on the people and activities in their sphere of control." But capitalists were capable of

[31]J. P. Nettl, "The State as a Conceptual Variable," *World Politics* 20 (July 1968), 559–92.

[32]This definition of the state is drawn from Weber, *Economy and Society*, 1:56, and Otto Hintze, "The State in Historical Perspective," in Reinhard Bendix, ed., *State and Society* (Berkeley: University of California Press, 1968), pp. 154–69. See also Theda Skocpol, *States and Social Revolutions* (New York: Cambridge University Press, 1979), esp. pp. 30–32.

movement (they had what Albert Hirschman calls "movable property"), and therefore state officials had to form alliances with various social classes and to foster capital accumulation. At the earliest moments in the building of European states, state leaders were confronted with a double-edged imperative: to harness domestic wealth so as to strengthen the state's foreign position, and to do so in a way that would not scare off capitalists or diminish economic growth.[33]

State intervention can stifle economic growth and diminish revenues even when mobility of productive resources is low. International competition, and the resulting need of the state to extract societal resources for military and foreign policy purposes, may constrain states even in relatively closed national economies. Indeed, the logic of a state's bargaining relationship with a society extends beyond considerations of economic growth. As Tilly notes, "the processes of bargaining with ordinary people for their acquiescence and their surrender of resources—money, goods, labor power—engaged the civilian managers of the state in establishing limits to state control, perimeters to state violence, and mechanisms for eliciting the consent of the subject population."[34] This underlying dilemma—the continuing need for the state to extract resources and impose costs on the society and economy while maintaining the confidence of economic actors and the consent of the population—is no less acute for the contemporary state. All states share these core features and face these basic domestic dilemmas, but they differ widely in the manner in which they are organized and the extent to which they can draw resources from and impose costs on the society and economy.

The domestic activities of the state cannot be separated from its

[33]Charles Tilly, "Warmaking and Statemaking as Organized Crime," in Evans, Rueschemeyer, and Skocpol, *Bringing the State Back In*, p. 172, and Albert Hirschman, "Exit, Voice, and the State," *World Politics* 31 (October 1978), 90–107. Douglas C. North spells out this fundamentally double-edged relationship between state and capital, presenting a more elaborate utility-maximizing model, in *Structure and Change in Economic History* (New York: Norton, 1981). See also Margaret Levi, "The Predatory Theory of Rule," *Politics and Society* 10 (1981), 431–65, and John A. C. Conybeare, "The Rent-Seeking State and Revenue Diversification," *World Politics* 35 (October 1982), 25–42. The state exchanges the provision of particular services (e.g., the maintenance of property rights) for privately generated revenue. It attempts to structure these services so as to maximize revenue. The state is constrained, however, since there is always the possibility that rivals can provide the same services. North argues: "Where there are no close substitutes, the existing ruler characteristically is a despot, a dictator, or an absolute monarch. The closer the substitutes, the fewer degrees of freedom the ruler possesses, and the greater the percentage of incremental income that will be retained by the constituents" (*Structure and Change*, p. 27).

[34]Charles Tilly, *Big Structures, Large Processes, Huge Comparisons* (New York: Russell Sage Foundation, 1985).

participation in the larger system of states. International geopolitical and economic pressures and opportunities, for example, can prompt states officials to attempt internal socioeconomic reforms or to direct the course of national economic development—activities that may or may not conflict with domestic social and economic interests. The ability of the state to protect or enhance the nation's international position hinges in important respects on the internal political and material resources of the nation and the state's access to them.[35]

Likewise, the state's international activities are intimately tied to its domestic position. The international system can provide an arena for state officials attempting to solve domestic problems, typically as states try to extract resources and political support from other states. Leaders of developing countries, for example, often find foreign aid and diplomatic recognition vital to the maintenance or enhancement of their domestic political standing.[36] Equally important, state officials may seek international agreements to redistribute or mitigate the economic and political costs of international economic change, thereby reducing the political and economic burdens those state officials would otherwise shoulder.[37]

A state's international position depends on its relative military and economic capabilities. These international capabilities refer to its access to and influence over the international system—access to and influence over the behavior of other actors. Similarly, the state's domestic position depends on its ability to perform a range of activities such as the reallocation of productive resources and the imposition of costs on the economy and society. Variations in these domestic and international capacities have implications for the manner in which states pursue adjustment policy.

The international position of a state influences its policy of adjustment. The more powerful the state is, the more it will emphasize international strategies. States that are strong internationally will attempt to solve adjustment problems within the international arena, thereby distributing the costs of adjustment among nations and re-

[35]This observation is emphasized by classical Realists. See Hans J. Morgenthau, *Power among Nations*, 6th ed. (New York: Knopf, 1985), and E. H. Carr, *The Twenty Years Crisis, 1919–1939: An Introduction to the Study of International Relations*, 2d ed. (London: Macmillan, 1962), chap. 8.

[36]See W. Howard Wriggins, *The Ruler's Imperative: Strategies for Political Survival in Asia and Africa* (New York: Columbia University Press, 1969).

[37]For a general discussion of international-domestic linkages see G. John Ikenberry, "The State and International Strategies of Adjustment," *World Politics* 39 (October 1986). See also Ikenberry, David A. Lake, and Michael Mastanduno, "Toward a Realist Theory of State Action," unpublished paper, 1986.

Table 2. Distribution of gross economic resources among the five major market-economy countries, 1960–1980

Year	United States	Germany	Britain	France	Japan	U.S. as % of Five Top Countries
1960	505	72	71	60	43	67
1970	989	105	122	141	204	60
1975	1539	420	141	339	499	51
1980	2587	819	204	652	1036	46

SOURCE. *United Nations Statistical Yearbook* (New York, 1981), p. 151.

ducing individual burdens. International strategies of adjustment usually require cooperation among nations. All states might benefit from an international strategy, but only powerful states can wield the political and economic resources necessary to gain agreement on their own terms and to induce others to accede to their wishes. In principle, combinations of weak states can agree to an international offensive strategy but, at the very least, they will find it difficult to do so. As a result, less powerful states are likely to emphasize domestic strategies.

The international position of a state can be gauged by a variety of measures. Indicators particularly relevant to the position of industrial countries relative to international petroleum markets are provided in Tables 2 and 3. The distribution of gross economic resources indicates that, despite a decline in relative position, the U.S. economy remained a clear international leader throughout the 1970s.

More revealing are measures of relative capability in the energy area. Although U.S. import dependence grew in the years following the 1973–74 crisis, imports remained a far smaller percentage of total energy consumption than in Europe and Japan. Aggregate imports were approximately half the size of European imports and roughly similar to those of Japan. The relative share of oil imports in domestic energy consumption was low, but the absolute share of imports in total imports by industrial countries was high, giving the United States greater opportunities than other consuming countries to exercise influence over international oil markets. Its large absolute share of imports gave the United States the ability to alter international supply and prices; the small place of those imports in domestic energy consumption meant that the costs of curtailing imports would be smaller for the United States, everything else being equal, than for other industrial importing countries.

These indicators suggest that in 1973, the United States was in a favorable international position to influence the course of the oil

Table 3. Petroleum imports of major Western countries, 1971–1979

a) Oil import dependence (oil imports as a percentage of total energy requirement)

	1971	1973	1975	1977	1979
United States	6%	10%	15%	20%	19%
France	67	75	64	65	64
Germany	42	42	37	44	39
Japan	68	74	70	70	66

SOURCE. Calculated from data in International Energy Agency, *Energy Balances of OECD Countries, 1971/1981,* Paris: OECD, 1983.

b) Oil imports (thousand barrels daily)

	1973	1975	1977	1979
United States	6,255	6,025	8,710	8,410
Western Europe	15,405	12,610	13,295	13,080
Japan	5,480	4,945	5,510	5,605

SOURCE. *BP Statistical Review of World Energy,* June 1984, p. 16.

c) U.S. oil balance

	Oil Imports as % of Consumption	Excess Production Capacity as % of Consumption	Net U.S. Position
1967	35	25	+6
1973	48	*	−25
1979	43	*	−40

*exact figure not available but close to zero.
SOURCE. Robert Keohane, *After Hegemony* (Princeton: Princeton University Press, 1985), p. 199.

crisis. However, the United States did not have surplus domestic production, a resource that had proved important in responding to earlier disruptions in international markets. To exercise its potential market power, therefore, the United States had to cut domestic consumption of imports. Unlike in previous supply disruptions, it could not simply release excess domestic production onto world markets. The substantial international capabilities of the American state thus hinged on its domestic capabilities; in this case, its ability to cut imports and adjust to higher oil prices.

The domestic position of the state influences the strategy of adjustment. The more constrained a state is by its relations with its economy and society, the more it will emphasize international strategies of adjustment. States that find it difficult to impose costs on their domestic societies will be more inclined to seek international solutions. Conversely, states that have the capabilities to redeploy domestic resources and impose the costs of change on society will emphasize domestic offensive strategies.

These relationships between the domestic and international capacities of states and adjustment strategies combine to provide a fuller set of propositions about state action. States that are internationally powerful but domestically constrained will have a dominant interest in an international strategy of adjustment; such was the position of the United States as it entered the oil crisis of 1973–74. U.S. executive officials pursued an international offensive strategy after 1973, although the nation's international power was insufficient for the strategy to succeed. States that are less powerful internationally, but that can redeploy domestic resources and impose costs on society, will emphasize domestic strategies of adjustment. The Europeans (the French most prominently) and the Japanese were in this position in 1973. None of these states was powerful internationally in the energy area, but each had various means to influence the course of domestic energy adjustment.

Disparities in international and domestic positions led the advanced industrial states to adopt initially divergent strategies of adjustment. I am *not* proposing a theory about the determinants of international collaboration. Rather, the advanced industrial states were confronted with divergent international and domestic constraints and opportunities, which influenced national preferences for specific forms of international agreements. In the American case, domestic constraints on adjustment (particularly oil price controls) made international options more attractive. At a later point those constraints made it difficult for the United States to play the international leadership role it envisaged. In contrast, the European and Japanese states had a wider array of options to pursue energy adjustment on a national or regional basis, and so to then the American proposals seemed less necessary.

The divergent positions of various states did not preclude international cooperation—indeed, some forms of cooperation emerged during the second oil shock. But the availability or absence of other options clearly influenced the value that states attached to cooperation. The failure of cooperative schemes forced U.S. officials to confront domestic capacities for adjustment, and in particular the organizational foundations of state capacities.

State Capacities and Alternative Adjustment Strategies

The role and capacities of executive officials are influenced by the organizational structures of the state in which they are situated. In the United States these structures are characterized by a fragmentation of political sovereignty across levels and branches of government and have left executive officials with a relatively modest bureaucratic planning capacity and few mechanisms for direct intervention in specific sectors. Organizational structures, however, are not immutable. At moments of national crisis, politicians and administrative officials have sought with varying degrees of success to expand their capacities to respond to economic crisis and change. Moreover, the capacities of executive officials should not be measured simply in terms of the mechanisms at their disposal. The ability to abstain from intervention or withdraw from specific sectors of the economy can also be crucial to the successful pursuit of policy objectives.

From conceptions of the capacity of executive officials that focus on the policy process and situate its characteristics within broader structures of state and society have emerged general propositions about the ability of leaders to accomplish state goals and about the weakness and strength of advanced industrial states.[38] A variety of organizational structures differentiate states according to their capacities. John Zysman, who is interested in the abilities of advanced capitalist states to become involved as coherent and weighty actors in industrial adjustment, focuses on mechanisms of recruitment in the national civil service, degree of centralization within government civil service, and independence of the civil service from legislative oversight. These characteristics of government structure, he argues, can be combined into a single measure of a state's capacity for intervention.[39]

The character of specific "policy networks" and "policy instruments," Peter Katzenstein argues, are shaped by the larger structures of state and society. The manner and effectiveness with which states can intervene in their economies are directly related to the character of their policy networks and the range of policy tools available to

[38]See Peter J. Katzenstein, "International Relations and Domestic Structures: Foreign Economic Policies of Advanced Industrial States," *International Organization* 30 (Winter 1976), 1–45; Katzenstein, *Between Power and Plenty;* Krasner, *Defending the National Interest,* esp. chap. 3; and Zysman, *Governments, Markets and Growth.*

[39]Zysman, *Governments, Markets and Growth,* p. 300. For discussions of the higher civil service see Ezra Suleiman, *Politics, Power and Bureaucracy: The Administrative Elite* (Princeton: Princeton University Press, 1974); John A. Armstrong, *The European Administrative Elite* (Princeton: Princeton University Press, 1973); Hugh Heclo, *A Government of Strangers* (Washington, D.C.: Brookings, 1977); and Suleiman, ed., *Bureaucrats and Policy Making* (New York: Holmes & Meier, 1984).

government officials: "The number and range of policy instruments emerge from the differentiation of state and society and the centralization of each." Each advanced industrial country represents a different balance between state and society.[40] Zysman is more concerned with the capacity of specific bureaucractic institutions, but both writers have notions of broad differences in the capabilities of states—differences derived from the historical play of political and economic development. State strength, they suggest, is fundamentally reflected in the organization and resources of bureaucratic government and its insulation from competing actors.

The most prominent strong state in these formulations is France, where permanent organizations of government officials have access to numerous policy tools and resources they can legitimately deploy in the regulation of a wide range of social and economic activities. There is every reason to place the state at the center of political analysis. Direct and indirect policy instruments can be used selectively to influence private economic decisions and practices. Moreover, networks of officials from private firms, state enterprises, and administrative bureaus reinforce the state's influence over the economy. French policy networks for trade, finance, and energy policy, Katzenstein finds, are state-centered, and American policy networks are society-centered.[41]

The United States is the exemplary weak state. The American state is fragmented and decentralized, possessing few of the instruments and resources necessary for strategic intervention in the economy and society. In no single issue-area is there a comprehensive, effective policy planning agency, much less for government as a whole, despite repeated presidential attempts to construct one. Frequent turnover among high-level executive officials and narrow divisions of bureaucratic authority create a "government of strangers" and severely hinder the emergence of effective bases of power.[42] The United States is not simply a late bloomer, but its political institutions are fundamentally malformed for the emergence of a strong state.

Useful in identifying variations in policy instruments and resources available to state officials, this approach helps us understand why U.S.

[40]Peter J. Katzenstein, "Conclusion: Domestic Structures and Strategies of Foreign Economic Policy," in Katzenstein, *Between Power and Plenty*, and Katzenstein, "International Relations and Domestic Structures," *International Organization* 30 (Winter 1976), 15.

[41]See Andrew Shonfield, *Modern Capitalism: The Changing Balance of Public and Private Power* (London: Oxford University Press, 1965), pp. 121–50; Suleiman, *Politics, Power and Bureaucracy;* and Katzenstein, "International Relations and Domestic Structures," p. 43.

[42]Heclo, *A Government of Strangers.*

officials had trouble with direct production of energy in the 1970s and soon turned to subsidies. However, strong and weak characterizations do not capture the nuances and variations in the capacities of states. The strong state of France has been challenged by descriptions of the role of intrest groups and clientilistic relationships more extensive than earlier depictions of a centralized, autonomous state recognized.[43] Recent studies of Japanese state-society relations also reveal a state bureaucracy severely constrained by interest groups and political parties.[44] Clearly state strength is more complex than a strong-weak distinction can capture.[45]

Likewise, the simple image of a weak American state, able to do little more than register interest group demands, is misleading. We need a more differentiated conception of state capacity, one that probes more deeply the fit between particular policy instruments and particular crises. It is precisely the search for such a fit that involved American executive officials in a sequence of adjustment proposals during the oil shocks of the seventies. In this sense, the capacity of the state must be discovered by the officials who seek to exercise it.

Moreover, state capacity is more than the expansion of state controls or activities. The withdrawal of direct regulatory controls or the abstension from intervention may also provide evidence of state capacity. Such considerations frustrate simple distinctions between strong and weak states and lead to an exploration of variations in state capacities within a single nation across time and issue-areas.

[43]In a new study of the French state and the notarial profession, Ezra Suleiman takes issue with the conventional, strong state image. Suleiman argues that the capacities of the French state are best understood in historically cyclical terms, where the interventionist efforts of a strong state to reshape societal groups result in transformations of relations with society and, ultimately, in a diminution of state capabilities. See *Private Power and Centralization in France: The Notaires and the State* (Princeton: Princeton University Press, 1988). See also Jack Hayward, *The State and the Market: Industrial Patriotism and Economic Intervention in France* (Brighton: Wheatsheaf, 1986); Harvey B. Feigenbaum, *The Politics of Public Enterprise: Oil and the French State* (Princeton: Princeton University Press, 1985); and Helen Milner, "Resisting the Protectionist Temptation: Industry and the Making of Trade Policy in France and the United States during the 1970s," *International Organization* 41 (Autumn 1987), 639–65.

[44]See the new interpretations of Japanese state capacities in Richard J. Samuels, *The Business of the Japanese State: Energy Markets in Comparative and Historical Perspective* (Ithaca: Cornell University Press, 1987), and David Friedman, *The Misunderstood Miracle: Industrial Development and Political Change in Japan* (Ithaca: Cornell University Press, 1988). The distinctive symbiotic relations between Japanese state and economy—what he calls "organized capitalism"—are also discussed by Ronald Dore in *Flexible Rigidities: Industrial Policy and Structural Adjustment in the Japanese Economy, 1970–80* (Stanford: Stanford University Press, 1986).

[45]See also G. John Ikenberry, "The Irony of State Strength: Comparative Responses to the Oil Shocks in the 1970s," *International Organization* 40 (Winter 1986), 105–37.

In the course of American energy adjustment, politicians and executive officials attempted to build additional state capacities for coping with the problems of energy adjustment. These efforts came on the heels of the failure of international cooperation. The early efforts at state building sought to increase the information about energy available to policy makers—this was the rationale of Project Independence. Later, a small group of officials proposed to create capacities for the direct financing of energy exploration and production, as in earlier crises seeking to enlarge the direct management ability of the state. These proposals failed, opposed by a wide array of interest groups and public officials within Congress and the executive branch. Some opponents feared the encroachment of the state on the private operation of energy finance and production. Others were skeptical about the purposes to which new state capacities would be put or about the possibility of their being subverted by private interests. At the same time the institutional legacy of the past weighed heavily on proposals for change. In particular, the scarcity and fragmentation of bureaucratic expertise and operational capability provided few bases from which to build new government powers and responsibilities. What state builders in the 1970s needed were organizational toeholds. Without them, state-building proposals were easily defeated.

Executive officials fell back on capacities or mechanisms that the American state has always favored. An enormous upsurge in spending made use of a traditional instrument of public policy, and a modest administrative organization was created to channel energy R & D funds into private hands. The spending option was attractive both to Congress and to the executive, particularly when financed from the proceeds of a tax on windfall oil profits. Senators and representatives could use the funds to nourish a wide assortment of constituencies; executive officials could use the programs to influence larger patterns of energy research and development. Nonetheless, research funding could not address immediate energy problems. Indeed, as these spending programs grew in size, they made the problems posed by price and allocation controls all the more apparent.

In the end executive officials discovered that the reconstruction of market pricing for petroleum was the most effective approach to energy adjustment. They faced formidable domestic obstacles. Petroleum price controls, enacted prior to the 1973–74 shock, partially shielded consumers and also advanced the commercial position of segments of the petroleum industry. These interests, represented in Congress, pushed successfully to preserve and extend price and allocation controls. Consequently, the move toward market pricing itself required vigorous state action.

The notion that the capacities of the state may be exercised in the withdrawl and not merely the extension of government controls is not appreciated by those who associate strength with direct state involvement in the economy and society. The analytical confusion lies in the dichotomy of state and market. The state (or authority) and the market (or exchange) are separate and distinctive modes of organization or social control, and, as Charles Lindblom argues, nations differ in the degree to which "market replaces government or government replaces market."[46] But differences in the balance of market and government activity are not in themselves indications of the strength or weakness of the state or of private enterprise. The political and economic ends that markets serve and the forces that organize and sustain those markets are empirical matters.

The state may organize or extend markets in the service of national and geopolitical objectives, of course, and the seminal statement of this view is Karl Polanyi's study of the nineteenth-century rise of English capitalism. Polanyi argues that the emergence of national markets was the result not of the "natural" and gradual expansion of nascent local and transnational trading relationships but rather of deliberate state policy. The construction of markets in England was the product of actions by a mercantilist state that saw national economic development as vital to strengthening the nation's international position.[47]

States also make more limited and sometimes very sophisticated efforts to enforce or extend market relations for a variety of political objectives. Japanese industrial policy, for example, occasionally uses as its central instrument the selective opening of domestic markets to international competitive pressures. State officials have used such market openings to encourage the rationalization of a domestic industrial sector; at other times they may wield the threat of market opening to encourage businesses to comply with other industrial policy goals. At the very least, the use of market mechanisms is likely to serve a mixture of private interests and state policy goals.[48]

In the United States the decision taken in 1979 to decontrol oil prices was not simply or primarily a ratification of private commercial interests. The policy itself was politically costly to the Carter admin-

---

[46]Lindblom, *Politics and Markets.*

[47]Karl Polanyi, *The Great Transformation* (New York: Farrar & Rinehart, 1944).

[48]Daniel Okimoto, *Between MITI and the Market* (Stanford: Stanford University Press, forthcoming), and John Zysman and Stephen S. Cohen, "The Mercantilist Challenge to the Liberal International Trade Order," paper prepared for the Joint Economic Committee of the U.S. Congress, 1982, p. 12. For a recent analysis of these relationships and processes of "mutual adjustment," see Michael Barzelay, *The Politicized Market Economy: Alcohol in Brazil's Energy Strategy* (Berkeley: University of California Press, 1986).

istration and engaged executive officials in an elaborate set of moves
to ensure its success. In preceding years, officials had pursued a variety of adjustment policies, involving efforts to transform the state's
role in the energy sector as well as programs to distribute massive
amounts of funds through existing government mechanisms. In the
end, however, executive officials found that their most powerful tool
was a return to the market.

CONCLUSION

My purpose is to explain why the industrial nations traveled divergent paths in their pursuit of energy adjustment and to explore the
implications of these divergent strategies for international cooperation among nations. I also seek to account for the evolution of energy
adjustment in the United States, which began with an international
strategy and ended with a decision to decontrol oil prices. A proper
understanding of American energy policy requires that we pay attention to the activity of executive officials as they are influenced and
constrained by larger international and domestic structures. In exploring the contours of energy adjustment, therefore, we are examining the larger logic of national adjustment to international change.

In this book I develop an institutional approach that advances several lines of analysis. First, I situate the organizational position of
executive officials in the 1970s in the larger historical trajectory of
state formation. Politicians and bureaucrats labor under the weight of
prior state building. The institutions in which these officials operate
and the legacies of previous policy inhibit opportunities for policy
action.

Second, I relate adjustment strategies to the broad international
and domestic position of states. Energy adjustment is part of a larger
set of challenges that arise in the course of international economic
change. The challenge of adjustment can be met through the pursuit
of either domestic or international policies, and in both domains policy can entail restructuring or protecting the existing order. At this
general level, the strategic choices made by executive officials depend
on state capacity; officials in a state that finds it difficult to impose
change on its own domestic system will emphasize international strategies. This logic of domestic and international strategies played itself
out in the early years of the 1973–74 crisis and helps explain the
failure of international cooperation on energy adjustment.

Finally, I develop my institutional approach by paying attention to the manner in which the organization structures of the state shaped the domestic sequence of American adjustment policies. Organizational structures of the state are crucial not only in which executive officials seek to accomplish but also in how they articulate what is at stake. They limited the types of policies executive officials could successfully pursue and help explain the sequence of energy adjustment strategies. This focus on executive officials shaping strategy but being constrained by the organizations they occupy orients my analysis.

CHAPTER THREE

The American State and Energy Policies

> Our energy resources are not inexhaustible, yet we are permitting waste in their use and production. In some instances, to achieve apparent economies today future generations will be forced to carry the burden of unnecessarily high costs and to substitute inferior fuels for particular purposes. National policies concerning these vital resources must recognize the availability of all of them; the location of each with respect to its markets; the costs of transporting them; the technological developments which will increase the efficiency of their production and use; and the relationships between the increased use of energy and the general economic development of the country.
>
> It is difficult in the long run to envisage a national coal policy, or a national petroleum policy, or a national waterpower policy without also in time a national resources policy.
>
> Franklin D. Roosevelt, 1939

Such a statement could have come from any embattled American president in the 1970s. It was actually made by President Roosevelt in February 1939, in a message to Congress transmitting the report and study of the National Resources Committe of Harold Ickes.[1] As FDR's message indicates, American public officials were concerned with energy supply and security issues long before 1973. Important episodes of government involvement punctuate the long history of international and domestic energy markets. The policy outcomes of these earlier struggles created the institutional terrain for energy adjustment policy in the 1970s.

[1]Franklin D. Roosevelt, *The Public Papers and Addresses of Franklin D. Roosevelt* (New York: Macmillan, 1941), February 16, 1939, p. 138.

In this chapter I sketch the origins and evolution of the American state—both its general organization and those structures related to energy policy making. Although state organization puts important constraints on the activities of executive officials, it is itself explicable in terms of distinctive phases of political and economic development. The pattern of energy adjustment policy in the 1970s can be understood only through a specification of the institutions and policy tools that accumulated in the fifty years prior to 1973.

ORGANIZATIONAL STRUCTURES OF THE STATE

The organizational structures of the American state were shaped by the grand forces of political and economic development. From a comparative perspective, the most powerful state bureaucracies are to be found in those advanced industrial nations where monarchical state building preceded the emergence of representative institutions.[2] Bureaucratic organization became one of the accessories of state making in Continential Europe. The building of states involved the territorial struggles of kings, the incorporation of existing political structures, the extraction of resources from a local population. State makers first had to raise taxes and deploy military forces, precisely the activities facilitated by bureaucratic staffs. Those political leaders in early modern Europe who developed specialized staffs and bureaucracies gained an "organizational advantage" over their rivals, and so national bureaucracy—relatively autonomous and powerful—attended the successful emergence of the first nation-states.[3]

American national political institutions, but contrast, were established without a strong or centralized state bureaucracy. The absence of an aggrandizing state maker and hostile external powers goes far to explain the modesty of the organizational center of the American state, which has remained "underdeveloped" because of the persistent prominence of other political mechanisms, such as political parties and the judiciary. As one author notes, "When the new nation was established in 1789, there was a very weak bureaucracy. And

[2]See Barrington Moore, Jr., *Social Origins of Dictatorship and Democracy: Lord and Peasant in the Making of the Modern World* (Boston: Beacon, 1966); Reinhard Bendix, *Kings and People: Power and the Mandate to Rule* (Berkeley: University of California Press, 1978); and Hans Rosenberg, *Bureaucracy, Aristocracy and Autocracy: The Prussian Experience, 1660–1815* (Boston: Beacon, 1958).

[3]So argues Charles Tilly in "Reflections on the History of European State-Making," in Tilly, *The Formation of National States in Western Europe* (Princeton: Princeton University Press, 1975), p. 30.

because of the decentralized federal system, political parties were essentially locally oriented. Indeed, throughout the nineteenth century, most of the federal bureaucracy was dominated by a locally oriented party system, thus reducing the ability of the central government to penetrate the society."[4]

A system of courts and parties, Stephen Skowronek argues, provided a powerful bulwark for political order in nineteenth-century America. In the last decades of the century a highly competitive national party system set constraints on administrative development. Electoral struggle was the dominant mechanism for the dispensation of political goods and power; politicians divided the spoils of office, and public policy fed the imperatives of state and local patronage systems. "The creation of more centralized, stable, and functionally specific institutional connections between state and society was impeded," as Skowronek notes, "by the tenacity of this highly mobilized, highly competitive, and locally oriented party democracy."[5] While political parties were emerging in European countries to challenge national administration, American reformers were seeking institutional alternatives to party dominance. State building in its American setting moved along a unique trajectory.

Much of the history of political reform around the turn of the century involved an effort to forge new bases for government authority. In its early phase, between 1877 and 1900, this reform struggle involved an ultimately unsuccessful attempt to establish a national administrative apparatus. Later, between 1900 and 1920, coalitions of reformers were able to break into national office and establish some forms of administrative power. These institutional breakthroughs, established within a rapidly growing and increasingly complex industrial order, brought bureaucratic administration into the political system. But the political strategies for reconstituting national political authority shifted with each presidential administration, and as a result bureaucratic administration remained dispersed and fragmented. The opportunities for state building were met with new forms of administrative authority, but in such a way as to encourage institutional disarray within the state.[6] The most important new form of administrative authority was the regulatory commission. This institu-

---

[4]J. Rogers Hollingsworth, "The United States," in Raymond Grew, ed., *Crises of Political Development in Europe and the United States* (Princeton: Princeton University Press, 1978), p. 165.

[5]Stephen Skowronek, *Building a New American State: The Expansion of National Administrative Capacities, 1877–1920* (New York: Cambridge University Press, 1982), pp. 39–40.

[6]Ibid., pp. 165–76.

tional creation provided a new center for public authority, but it did so at the expense of centralized bureaucratic capacities.

The modern growth of executive administration has seen a continued layering and specialization of government bureaucracy. The postwar expansion of policy issues has greatly enlarged the scope of the federal bureaucracy but has further attenuated presidential control of administration. The growth in administration, Hugh Heclo observes, has occurred at the middle level of government, where narrow professional experts, organized in webs of associations that blur public and private boundaries, have gained increasing prominence and autonomy. The lines of power are arranged in complex "issue networks" that cut across government and societal organizational structures: "Participants move in and out of networks constantly. Rather than groups united in dominance over a program, no one, as far as one can tell, is in control of the policies and issues."[7]

The American federal organization is distinctive because of its fragmentary, shifting, and narrowly specialized institutions. Bureaucracy within the executive establishment has grown, yet its authority and capacities have remained strikingly delimited. Obviously these persistent institutional conditions inhibit broad-scale policy planning. Bureaucratic organization remains a matter of small units, and administrative authority has been built on the basis of short-term political control of specialized expertise that is not attached to bureaucratic organization. Programmatic policy planning and the capacity to bring that policy into decision-making realms remain elusive in such a context.

This sketch suggests the American state is a continually transformed, internally differentiated area of conflict rather than an integrated institution. Andrew Shonfield has already noted the implication of the U.S. administrative structure—a "loose confederation of more or less hostile bodies"—for policy coherence. Whereas in other Western nations, "the general trend is to use the aggregation of public power in order to create a coherent force whose significance will be greater than the sum of its individual parts," in the United States, a "curious disorder" resides at the "heart of the American administrative process." Skowronek depicts the modern American administrative state as a "hapless giant," a state that could "spawn bureaucratic goods and services but that defied authoritative control and

[7]Hugh Heclo, "Issue Networks and the Executive Establishment," in Anthony King, ed., *The New American Political System* (Washington, D.C.: American Enterprise Institute, 1978), pp. 87–124, quotation at p. 102.

direction."[8] Such patterns of state organization can be found in government-business relations as well.

Patterns of Government-Business Relationships

The distinctive political development of the nineteenth century gave shape to an American state in which a centralized and weighty bureaucratic core was striking by its absence. That sequence of political development, when juxtaposed with the phases of economic development, also provides a powerful explanation for patterns of state involvement in the economy. Early industrializers did not face an established, competitive world industrial order, nor were the first industrial sectors, such as textiles, as capital-intensive as the later sectors. Accordingly, private entreprenuers did not have the same reliance on government for investment needs. As new industries emerged, moreover, already accumulated capital kept investment resources and industrial decision making beyond the realm of state control.

Those nations which industrialized relatively late ("backward countries") faced larger and more complex problems of economic development.[9] The types of industries that backward nations needed to encourage, such as steel, required large-scale and intensive industrial, banking, and state organizations. Large investment and export promotion served to bring the government into a close relationship with industry, and industry's economic dependence on government brought political dependence with it. In late-industrializing Germany, for instance, relations between banks and industry were close. The investment needs of German industrialists, requiring rapid and massive capital expenditures, differed substantially from those of Britain. As Alexander Gerschenkron argues,

> the industrialization of England had proceeded without any substantial utilization of banking for long-term investment purposes. The more gradual character of the industrializing process and the more considerable accumulation of capital, first from earnings in trade and modernized agriculture and later from industry itself, obviated the pressure for

[8]Andrew Shonfield, *Modern Capitalism: The Changing Balance of Public and Private Power* (London: Oxford University Press, 1965), pp. 318–19; Skowronek, *Building a New American State,* p. 290.
[9]Alexander Gerschenkron, "Economic Backwardness in Historical Perspective," in Gerschenkron, *Economic Backwardness in Historical Perspective: A Book of Essays* (Cambridge: Harvard University Press, 1962), pp. 5–30.

developing any special institutional devices for provision of long-term capital to industry. By contrast, in a relatively backward country capital is scarce and diffused, the distrust of industrial activities is considerable, and, finally, there is greater pressure for bigness because of the scope of the industrialization movement, the larger average size of plant, and the concentration of industrialization processes on branches of relatively high rates of capital output.[10]

Late industrializers will have more concentrated financial and industrial institutions, and the state is likely to be more directly involved in industrial development. State dominance does not simply reflect the level of economic backwardness, however, for complex historical antecedents conditioned the state's involvement. For example, the Russian state, without large-scale banking institutions, sought rather unsuccessfully to use direct government intervention in order to force industrialization. German state involvement, on the other hand, was directed at infrastructural investment and the maintenance of political stability.[11]

It was not simply the demands of national economic welfare or efficiency that led the state directly into the economy of late-industrializing countries. A geopolitical logic made national defense important to European state involvement in the economy. As one analyst notes, "only the perception of external threat from the prior presence of already industrialized countries prompted considerably enhanced State involvement with industrialization, and explains the rapidity of progress."[12] Geopolitical pressures created incentives for the state to speed up industrialization and encourage industries of strategic importance to the national economy. This geopolitical logic and the unique capital and technological needs of late-arriving industries gave the state an added presence in the economy.

[10]Ibid., p. 14. The argument has been developed by James R. Kurth, "The Political Consequences of the Product Cycle: Industrial History and Political Outcomes," *International Organization* 33 (Winter 1979), 1–34, and Kurth, "Industrial Change and Political Change: A European Perspective," in David Collier, ed., *The New Authoritariansim in Latin America* (Princeton: Princeton University Press, 1979). The theory has been extended to Latin America also. See Albert O. Hirschman, "The Political Economy of Import-Substituting Industrialization in Latin America," *Quarterly Journal of Economics* 82 (February 1968), 2–32, and Philippe C. Schmitter, "Paths to Political Development in Latin America," in Douglas A. Chalmers, ed., *Changing Latin America: New Interpretations of Its Politics and Sociology* (New York: Academy of Political Science, Columbia University, 1972), pp. 83–105.
[11]Gerschenkron, "Economic Backwardness," pp. 16–21. See further Peter Gourevitch, "The Second Image Reversed," *International Organization* 32 (Autumn 1978), 881–912.
[12]Gautam Sen, *The Military Origins of Industrialisation and International Trade Rivalry* (New York: St. Martin's, 1984), p. 79.

In the United States the early growth and maturation of big business and the strikingly late growth of national public administrative institutions were, Alfred Chandler argues, crucial for the distinctive American relationship between government and business.[13] Unlike in Continental Europe, where a state bureaucratic apparatus appeared before large-scale industrialization, in America business growth preceded national institutions. In the crucial early period of industrial development the federal government remained outside the emerging institutional realms of economy and society.

Also of importance to American business-government relations, according to Chandler, was the substance of economic growth. External trade was of early importance to European industry, which reinforced collaborative relations with the state by creating incentives to seek government assistance in securing overseas markets. Furthermore, the emerging large firms in Europe tended to be in heavy industry, such as chemicals, metals, and machine tools, and did not have smaller entrepreneurial rivals. In the United States, however, the demands for government involvement were quite different. With large domestic markets and intense conflicts between large business and small merchants and wholesalers, the state played a peacekeeping rather than a promotional role.

Internal conflicts in American business and the limited organizational capacities of the state made antitrust and regulatory tools the state's dominant means of intervening in the turn-of-the-century national economy. Regulatory institutions grew in reaction to rapid growth of large business, first the railroads and later industrial manufacturers, and in a context shaped by contention over vertical integration and intrabusiness hostilities. Between 1887 and 1914 an institutional framework was established through the Interstate Commerce Act, the Sherman Anti-Trust Act, and the Federal Trade Commission Acts. As Chandler argues, "when the large government bureaucracies did come in this country, the basic role of government toward business had already been defined; and that definition developed largely as a response to an influential segment of business community to the rise of modern big business. A comparable response did not occur abroad."[14]

[13]Alfred D. Chandler, Jr., "Government versus Business: An American Phenomenon," in John T. Dunlop, ed., *Business and Public Policy* (Cambridge: Harvard University Press, 1980); also Chandler, *The Visible Hand: The Managerial Revolution in American Business* (Cambridge: Harvard University Press, 1978); and Chandler, "The Coming of Big Business," in C. Vann Woodward, ed., *The Comparative Approach to American History* (New York: Basic Books, 1968), pp. 220–35.

[14]Chandler, "Government versus Business," pp. 4–5, 10–11.

Regulation in Comparative Perspective

The unique centrality of regulatory mechanisms to American government-business relations reveals much about the conjunctures of political and economic life at a crucial stage of industrial development. One school of thought argues that American regulatory bodies have advanced "public interest" claims in the face of uncompetitive business practices. Others have emphasized the "capture" of regulatory agencies by business itself or, more strongly, claim that regulatory mechanisms emerged as instruments of industrial interests.[15] The complexity of the regulatory process, the variety of agencies and industries involved, and the long expanse of regulatory history inhibit a settling of this argument.[16]

It is clear that regulatory institutions emerged in response to industrial and political conflict, usually conflict within a particular industry. Furthermore, the important issue that informed public, legal, and government debate—and the conflict that called for a public response—was the emergence of the large, integrated business enterprise. Between 1897 and 1900, 183 large industrial firms were formed, commanding one-seventh of all industrial manufacturing capacity.[17] Unlike European states, which in different geopolitical and economic settings actively sought to encourage cartelization of major leading industries, the American state became involved in the economy because of internal business conflicts over size and competitiveness.

Regulatory bodies were established in response to pressures from clusters of vocal small businessmen, progressive reformers, and politicians using antimonopoly and procompetitive ideological appeals. The style of regulation evolved in the early decades of the twentieth century. The first industries subject to regulation were capital goods industries: railroads, steel, oil, coal. After World War I regulation expanded to include consumer goods: automobiles, appliances, packaged foods. The evolution of regulatory control followed a stream of

[15]For an overview, see Thomas K. McCraw, "Regulation in America: A Review Article," *Business History Review* 49 (Summer 1975), 159–83. The classic statement of this public interest position is James Macauley Landis, *The Administrative Process* (New Haven: Yale University Press, 1938). On capture see especially Gabriel Kolko, *The Triumph of Conservatism: A Reinterpretation of American History, 1900–1916* (Chicago: Quadrangle, 1963).

[16]See James Q. Wilson, ed., *The Politics of Regulation* (New York: Basic Books, 1980).

[17]See Thomas K. McCraw, "Regulatory Agencies," in Glenn Porter, ed., *Encyclopedia of American Economic History* (New York: Scribner's, 1980), p. 788, and Morton Keller, "The Pluralist State: American Economic Regulation in Comparative Perspective, 1900–1930," in Thomas K. McCraw, ed., *Regulation in Perspective* (Cambridge: Harvard University Press, 1981), p. 68.

Supreme Court decisions that delinated the proper sphere of regulatory action in setting standards and rates, in market conditions, and in antitrust.[18]

The regulatory approach was particularly compelling in the United States. In a rapidly expanding economy with large internal markets, the government was not confronted directly with a need to cartelize basic industry for purposes of international competition. Rather, its central task was the incorporation of public interest goals into conflicts among businesses and between business and society. Strong collaborative relationships between business and government were not necessary.

Just as important, the slowly evolving set of federal and state-level political institutions limited the opportunities for government intervention. The regulatory system created new realms of public authority, based on regulatory commissions staffed by experts and with the competence to apply administrative law as defined by the American judiciary. Institutional structures were created to bring government and business together, but the relationship remained at arms' length, delimited by the courts, and substantially beyond direct congressional or bureaucratic reach.[19]

The regulatory mechanism was a modest tool of government, and around it reformers and capitalists frequently struggled. This form of articulation between government and business has retained a durability that stems partly from its flexibility. New Deal policy innovations, for instance, were built upon established regulatory controls: temporary industrial commissions were established to supervise regulated industries and to adjust the combinations of public and private officials charged with administering guidelines. But this adaptability of institutional form has also had a stunting effect on the competence and capacity of central government bureaucracies. The hallmark of the regulatory mode is dispersed centers of government competence, deployed in a reactive manner and directed at defining market shares and competitive practices. The mode does not include organizational incentives for added levels of planning and programmatic government policy.

When the institutional structures defining business and government relationships were formed, American government was left with

---

[18]See Keller, "The Pluralist State," pp. 74–94.

[19]This logic can be found in social policy as well. See Theda Skocpol and John Ikenberry, "The Political Formation of the American Welfare State in Historical and Comparative Perspective," *Comparative Social Research* 6 (Greenwich, Conn.: JAI Press, 1983), 359–81.

primarily reactive, peace-keeping responsibilities. The prevailing organizational structures of the state positioned public authority in peripheral regulatory commissions.[20]

Macroeconomic Governance

Other methods of government intervention evolved in the 1930s and after, but the types of policy tools that emerged remained relatively indirect and undifferentiated by sector or industry. Policy was directed at the influencing of aggregate economic performance through gross fiscal expenditures and monetary policy.[21]

In 1968 the American delegation to the OECD explained the importance of what it called "global steering" as industrial policy: "The United States Government has come to recognize to a greater extent than ever before that its objective of steady economic growth without significant inflation requires it to co-ordinate its programmes with respect to Government spending, taxation and Federal Reserve discount rates and reserve requirements. . . When the fiscal and monetary policies of the Government are co-ordinated they constitute very powerful tools for providing systematic guidance to the economy as a whole." The report noted that American government structure did not allow for systematic and industry-specific interventions. "The Federal administrative structure is not designed to carry out an active, co-ordinated policy of promoting industrial growth. . . Intervention has normally been *ad hoc*. Our actions are developed as the problem and the occasion arise: the structure for dealing with such matters is thus a reactive one rather than a formal planning structure."[22]

These aggregate economic measures allowed the federal government to assume control of the economy without altering the structures—private, corporate decision making and arms-length public-private relations—themselves. This was the genius of Keynesian eco-

[20]As one analyst notes: "While most other states in capitalist societies increased their role and power as industrialization proceeded, the authority of the American state declined and its size remained relatively small. When seen in comparative terms, it simply had a less necessary role to play." David Vogel, "Why Businessmen Distrust Their State: The Political Consciousness of American Corporate Executives," *British Journal of Political Science* 8 (1978), 57. Also see Thomas K. McCraw, "Business and Government: The Origins of the Adversary Relationship," *California Management Review* 26 (Winter 1984), 33–52.

[21]See Herbert Stein, *The Fiscal Revolution in America* (Chicago: University of Chicago Press, 1969).

[22]*United States Industrial Policies*, Observations presented by the U.S. Delegation before the Industry Committee of the OECD at its 6th Session (Paris: OECD, 1968), pp. 27, 35.

nomics, for within the prevailing capitalist system government officials could fashion fiscal and monetary policy to influence aggregate spending, thereby influencing the investment climate and incentives for economic growth. Only modest levels of government autonomy were needed for this form of economic control. And among the variants of Keynesian economics, the least interventionist forms were incorporated into government activity.[23]

Since the 1930s American government has developed the macroeconomic tools and a modest planning apparatus to make "global" economic policy. These new economic tools were built upon institutions that did not involve direct and selective interventions. The result was a style of government activity that conformed, in Hugh Heclo's phrase, to the "idea of government by remote control." Heclo argues that "political administration in Washington is heavily conditioned by an accumulation of methods for paying bills and regulating the conduct of intermediary organizations. This pattern is consistent with a long tradition of fragmented and decentralized administration."[24]

THE EMERGENCE OF THE PETROLEUM INDUSTRY

The historical shaping of the American state and the patterns of government-business relations are also revealed in the petroleum business. The timing and phases of industrial development and state building are again crucial to an understanding of the larger historical patterns. The petroleum industry had become a major domestic and international industry *before* the state actively became concerned with national energy goals. When the state did get involved, it was primarily in a reactive fashion, addressing domestic stability and security of international supply. The federal government did not develop a tutelary relationship with the industry, as occurred in other Western countries during the early decades of the twentieth century. Regulatory mechanisms typified domestic government intervention, and diplomatic initiative typified government involvement in the foreign operations of American-based international petroleum companies.

[23]See Robert Skidelsky, "The Decline of Keynesian Politics," in Colin Crouch, ed., *State and Economy in Contemporary Capitalism* (New York: St. Martin's, 1979), pp. 62–63; Robert Lekachman, *The Age of Keynes* (New York: Random House, 1966); Robert M. Collins, *The Business Response to Keynes, 1929–1964* (New York: Columbia University Press, 1981); and Margaret Weir and Theda Skocpol, "State Structures and the Possibilities for 'Keynesian' Responses to the Great Depression in Sweden, Britain, and the United States," in Peter B. Evans, Dietrich Rueschemeyer, and Skocpol, *Bringing the State Back In* (New York: Cambridge University Press, 1985), pp. 107–63.

[24]Heclo, "Issue Networks," p. 92.

Early Growth of the Petroleum Industry

The American petroleum industry grew into a multinational system in the years following the turn of the century. The international growth of this American-centered industry accompanied its evolution into an oligopoly of firms, large in size and small in number, which divided and controlled production and distribution.

The early modern discoveries of oil were in the United States and in Russian Caucasia. As these sources of petroleum were developed, and the economies of the industrialized world generated further demand, a multitude of firms emerged to produce, refine, and market the new source of energy. In the United States the early years of the industry were characterized by intense competition among a large number of firms. The original sources of oil were in Pennsylvania and Ohio; markets were concentrated in New England, the Middle West, and Europe. By the turn of the century discoveries had been made in the Dutch East Indies, Burma, and Venezuela, and in this period individual companies could secure the whole of a producing country's territory as a concession.[25]

Very early in the development of the industry, however, large integrated firms came to dominate. Characteristics unique to the industry, involving economies of scale, geographical logistics of source and markets, and finite sources of supply, conditioned this concentration of industrial power. The story of the Rockefeller empire at Standard Oil has been told many times.[26] Prior to the restraint of trade action of 1911 and the breakup of the empire, the Standard Oil company used its market power to consolidate a staggeringly large position within the industry. The monopoly position of the American industry changed in the years after World War I, owing both to the breakup of Standard Oil and the emergence of smaller competitors in Texas. The important point here, however, is that the American industry emerged under conditions that did not require the state to play an active role in its early growth.

Yet the American state did come to play a role in the years between the world wars, when oil firms were expanding and stabilizing their operations. Several industry issues brought firms into the policy pro-

[25]See Harold F. Williamson and Arnold R. Daum, *The American Petroleum Industry: The Age of Illumination* (Evanston: Northwestern University Press, 1959), p. 320, and Evan Luard, *The Management of the World Economy* (New York: St. Martin's, 1983), p. 143.

[26]See Ed Shaffer, *The United States and the Control of World Oil* (New York: St. Martin's, 1983), pp. 20–22. On Standard Oil see, for example, Ira M. Tarbell, *The History of the Standard Oil Company*, 2 vols. (New York: Macmillan, 1933), and Harold F. Williamson, Ralph L. Andreano, Arnold R. Daum, and Gilbert Klose, *The American Petroleum Industry: The Age of Energy, 1899–1959* (Evanston: Northwestern University Press, 1963).

cess. Independent firms sought government regulation to stabilize the wild fluctuations in production and prices. Others sought antitrust action to break up the dominant majors. International firms sought government action to gain access to new foreign oil fields.

At the same time executive officials were concerned with two major issues that complemented some of the concerns of American energy firms. Government officials, wanting to encourage open and relatively liberal markets in the sector, used regulatory and antitrust instruments. Also, federal government officials perceived at various junctures a need to encourage the expansion of American firms into foreign oil fields. Here diplomatic assistance was a key instrument. In Europe, by contrast, the patterns of state promotion in energy industries were somewhat different.

Foreign Government Involvement in the International Oil System

In the years before World War I the European powers actively encouraged their national firms to develop operations in promising areas of petroleum production. Britain was an early leader. Winston Churchill, among others, argued that oil was superior to coal as a fuel for battleships, and with war in the offing he advocated government involvement in Persian oil fields through the creation of a "national champion" company. In 1914 the British Parliament, by a vote of 254 to 18, agreed to buy controlling shares in a petroleum firm, and thus the British government became partowner in the Anglo-Persian Oil Company.[27]

State involvement was understood to provide financial and political support for company operations. In return the contract stipulated that Anglo-Persian would remain a British concern, with the government appointing two of the seven company directors. This decision to acquire a private petroleum firm became in many respects a model for subsequent governments. As one analyst notes, "it was a prototype. Britain sought and gained command of secured 'tied' supplies of oil from a promising source that later turned out to be part of the richest oil-bearing region yet discovered."[28]

Commercial and government activity quickened with the end of

[27]Louis Turner, *Oil Companies in the International System* (London: Allen & Unwin, 1979), p. 25; Raymond Vernon, "Enterprise and Government in Western Europe," in Vernon, ed., *Big Business and the State: Changing Relations in Western Europe* (London: Macmillan, 1974), p. 11; Christopher Tugendhat and Adrian Hamilton, *Oil—The Biggest Business* (London: Eyre Methuen, 1975), p. 68.

[28]J. E. Hartshorn, *Oil Companies and Governments: An Account of the International Oil Industry in Its Political Environment* (London: Faber & Faber, 1962), p. 233.

World War I, in the creation of an international petroleum system from the relics of the Ottoman Empire. European governments actively involved themselves in securing access to Middle Eastern territories. In 1920, for example, it was agreed among the parties that Britain would have control over Mesopotamia and the French would assume the erstwhile German interests in the Turkish Petroleum Company. France, which had been desperately dependent on foreign oil sources during the war, moved to create its own petroleum firm, the Compagnie Française des Pétroles (CFP). CFP, partially owned by the government, represented French interests in Iraqi and other Middle Eastern oil fields.[29] In 1928 the French government enacted extensive regulations in order to encourage the construction of independent national refining capacity.

By the late 1920s a few international companies had come to dominate the international production and trade in oil. The largest of these were the majors, also known as the "Seven Sisters": Jersey Standard (later Esso and Exxon), Shell, Anglo-Iranian (later BP), Socony (later Mobil), Socal, Texaco, and Gulf.[30] The large companies were involved in all of the industry stages—extraction, refining, and marketing. Some had developed their own extensive supplies of crude. Gulf had supplies in the United States, Mexico, and Venezuela, and later Kuwait and Nigeria; Socal had supplies in California, Bahrain, and eventually Saudi Arabia. Others, such as Jersey Standard, Shell, and Socony, did not have extensive supplies and had to buy from others to meet their refining and marketing needs.[31] As the market for petroleum continued to grow, the various companies expanded their foreign exploration and acquired new concessions. Much of this expansion occurred in the Middle Eastern countries of Saudi Arabia, Iran, Iraq, and Muscat and Oman.

A renewal of state concern about oil security and development followed World War II, and again governments became directly involved in the petroleum industry. The French government, for instance, created two public firms—the Bureau de Recherches de Pétrole (BRP) and the Régie Autonome des Pétroles (RAP)—mandated to explore for and produce petroleum. These new French undertakings were pushed forward by a senior French official, Pierre Guillaumat, who had strong ties among political, civil servant, and business groups. Guillaumat, who was directeur des carburants from

[29]See Turner, *Oil Companies*, pp. 26–27.

[30]For background see Anthony Sampson, *The Seven Sisters: The Great Oil Companies and the World They Shaped* (New York: Bantam, 1975).

[31]Luard, *Management of the World Economy*, p. 145.

1944 to 1951, also became director of BRP in 1945. Later, in the 1950s, he would be instrumental in promoting the French civilian nuclear power industry. Moving from one top post to another in government and in oil and nuclear enterprises, Guillaumat reflected in his career the institutional possibilities for collaborative relations among public and private organizations.[32]

The French state's postwar push into the petroleum industry was matched elsewhere. Italy and Norway also created new public enterprises or nationalized old ones. Several factors shaped the state's role in the oil industry. The most important determinant of initial government involvement was the magnitude of national dependence on foreign oil. France, more than Britain and Germany, for example, depended heavily on imported oil that was outside the control of French firms. Britain and Germany relied much longer than France on coal to supply a substantial portion of national energy needs. The pattern of involvement was further shaped by prevailing ideological and political conceptions of the proper role for public institutions to play in business activity—France and Italy had already developed an extensive government presence in business development. Finally, the existence of established, large multinational firms gave certain countries—notably the United States, the Netherlands, and Britain—less reason for direct government involvement.[33] Thus challenges of resource dependence combined with national objectives of political control to mold separate styles of state involvement in the petroleum industry. After World War II, French, Italian, German, and Japanese government renewed their individual efforts to gain some measure of national autonomy in an international petroleum system dominated by the Americans and the British.

Nonetheless, between the late 1920s and the late 1960s a remarkable stability was maintained within the international oil regime. Large, integrated companies extended and stabilized their sources of crude petroleum, and predictable marketing systems were developed. There was little competition between the major firms. A mutuality of interest between these companies allowed a sophisticated system of oil sharing, even collusion, to emerge. Competition, to the degree that it existed, was primarily focused on gaining access to supplies. This managed aspect of the international oil regime was epitomized in the so-called Red Line agreement of 1928, in which five of the multina-

[32]See Oystein Noreng, "State-Owned Oil Companies: Western Europe," in Raymond Vernon and Yair Aharoni, eds., *State-Owned Enterprise in the Western Economies* (New York: St. Martin's, 1981), pp. 133-34, and Turner, *Oil Companies*, p. 57.
[33]See Leslie E. Grayson, *National Oil Companies* (New York: Wiley, 1981), p. 7.

tional companies involved in the Iraq Petroleum Company operation agreed not to seek further concessions in territories of the former Ottoman Empire. Under the "as is" agreement of the same year, moreover, the largest of the Seven Sisters agreed to specific pricing and market-sharing agreements.[34]

Later, combinations of integrated firms produced agreements jointly to exploit newly discovered reserves in Kuwait, Saudi Arabia, and elsewhere. The most extensive agreement came when five large firms jointly established Aramco to develop the massive Saudi Arabian concession. The Red Line agreement was later challenged by the American government in order to allow U.S. company access to the area, but these various agreements revealed the essentially negotiated nature of the newly established international petroleum regime.

American Government Involvement in the International Petroleum System

State officials became actively involved in supporting access to sources of oil for their national companies after 1919. The federal government in the United States, in contrast to the British and French governments, which favored direct corporate involvement, remained largely a source of diplomatic support. The United States by 1914 had a very large domestic petroleum industry, and problems of supply or security were matters of at most distant concern. Indeed, the first government involvement in the activities of American firms was the effort, culminating in 1911, to break up Standard Oil. Nonetheless, American companies did seek support from the State Department to help gain access to the territories of the former Ottoman Empire. U.S. government involvement took the form of diplomatic protests, primarily directed at the British, arguing for "open door" commerical conduct. In the end this pressure succeeded in convincing British, French, and Dutch interests to allow American corporate involvement in the Turkish Petroleum Company and in Mesopotamian oil.

During the 1920s, despite postwar concerns about oil supply and security, the American state did not adopt the national champion approach. Five large American companies with foreign operations had already emerged, making direct government involvement through a national champion more problematic. After World War I government concern developed about the depletion of domestic reserves, leading at least one senator (Phelan of California) to call for the creation of a government-owned corporation, yet a sequence of

[34]Sampson, *Seven Sisters*, pp. 70–103.

Republican administrations in the 1920s made such a notion unlikely. Finally, the commanding American position in world oil production, which was as high as 80 percent in 1920, gave little force to calls for greater government involvement.[35]

International and Domestic Firms

Although involvement by the American state was limited to diplomatic initiatives within the emerging international petroleum regime, the domestic industry provided a basis for regulatory intervention. Other industrial nations had modest domestic resources; The United States, on the other hand, had massive petroleum reserves. This difference created the basis very early in the industry's history for a diversified American industry; diversified between producers, refiners, and marketers, and diversified in terms of large producers and small, independent operations. This latter division was particularly important in generating pressures for reactive government intervention.

The changing structure of the American petroleum industry after World War I, as new competitors emerged, stemmed in part from the breakup of Standard Oil. The energy problems perceived by American public officials and politicians were not as straightforward as problems were in Europe. Whereas the initial policy concerns for European governments involved access to and security of energy supplies, American government officials initially worried about the looming size of the Standard Oil empire. The government's posture toward to oil industry followed a general, national pattern: the use of regulatory and antitrust policy tools to restrain big business and engage in industrial peacekeeping.

The state's concern about the oil trust was buttressed by the presence of threatened small, independent firms. Independent oil firms occupied niches within the larger petroleum industry. These disadvantaged interest—local producers, refiners, and distributors—maintained pressure on government in its protracted antitrust deliberations.[36] The niches they occupied were created in part, ironically, from Rockefeller's early corporate strategy. The Standard Oil Company made a corporate decision not to invest in new domestic sources of oil. It aimed, rather, to control the industry through dominance of downstream refining and marketing facilities. The discovery of large

[35]See Turner, *Oil Companies*, pp. 27–28.
[36]Keller, "Pluralist State," p. 68.

oil fields in Texas in 1901 allowed new competitors to enter the industry. In the year Standard Oil was broken up, the company owned only about 10 percent of the new Texas production. As one analyst notes, "the new entrants, whose power was based on the ownership of crude rather than refining capacity, were able to breach Rockefeller's domestic monopoly even before the Supreme Court acted."[37]

The influence of the independents was (and still is) substantial. As Raymond Vernon notes, "the strength of the independents, then as now, rested in part on the fact that they were well distributed over the face of the United States and could rally formidable Congressional support for any position they took. The ability of the independents to provide a source of countervailing power would appear again and again in the shaping of U.S. oil policy during the decades to follow."[38] The organizational structure of the state enlarged the leverage of such companies over policy making, and beyond what economic clout alone might explain.

Following the early period, when antitrust was the central tool, executive officials became more concerned with influencing the management of an increasingly sprawling and disparate industry.[39] In the midst of the diversity of petroleum interests, and with an influential domestic industry, the state adopted a style of involvement that emphasized tax and regulatory intervention. Special tax provisions began in 1918 to encourage the development of coal, gas, and petroleum resources. In the 1930s the central problem for domestic petroleum producers was the overproduction of domestic oil, which led to a state-level regulatory response—"market-demand prorationing"—that acted to restrict production and preserve market shares.[40]

Federal government interest in regulatory intervention focused on the conservation of supplies, the maintenance of minimal levels of competition, and the insurance of secure sources of oil for military purposes.[41] Federal regulation of the natural gas market began in the 1950s. The resulting regulatory institutions were a fragmented mixture of decentralized and improvised mechanisms that left govern-

[37]Shaffer, *United States,* p. 30.

[38]Raymond Vernon, "The Influence of the U.S. Government upon Multinational Enterprises: The Case of Oil," in *The New Petroleum Order: From the Transnational Company to Relations between Governments* (Quebec: Les Presses de l'université Laval, 1976), p. 56.

[39]Gerald D. Nash, *United States Oil Policy, 1890–1964* (Pittsburgh: University of Pittsburgh Press, 1968), chap. 1.

[40]Cf. Arthur W. Wright, "The Case of the United States: Energy as a Political Good," *Journal of Comparative Economics* 2 (1978), 168–69.

[41]The narrative sketch of the next few pages is drawn from several sources; especially useful is Nash, *United States Oil Policy.* The quotation is from Nash, p. 120.

ment at the periphery of the industry and the conflicts between various industry groups unsettled.

During World War I, and at other moments of intense government concern over oil security and supply, the emphasis of federal petroleum policy shifted. Rather than try to influence the structure and competitive status of the industry, executive officials sought to promote cooperative government-business relations and to make regulatory policy more effective. These new attitudes toward the petroleum industry, which would characterize government policy well into World War II, were developed in the Fuel Administration of Woodrow Wilson's government. With oil conservation and production issues attracting high-level government attention during World War I, the Fuel Administration advanced regulatory policy that would further integrate the industry. This style of administration would become a basic feature of government intervention in the petroleum industry during times of crisis: the government promoted self-regulation by industry leaders. Seeking to stabilize production, for example, the Fuel Administration organized producers from various regions of the country to coordinate production levels. Because of the rising wartime need for petroleum, moreover, politicians and administrators advocated tax reductions for domestic producers. A consensus developed between government and oil producers—government would give the industry favorable tax treatment, and in return the industry would respond to the heightened petroleum needs associated with national defense.

After wartime concern over production and security of supply waned, the federal government's influence dwindled, and industry leaders held on to their newfound tax and regulatory prerogatives while seeking to reestablish erstwhile levels of business autonomy. In the late 1920s new conflicts broke out between independent producers and international firms, and the first major conflicts appeared between domestic independents and oil importers. The position favored by President Herbert Hoover involved voluntary intraindustry cooperation to restrain imports. Later, during the Depression, when oil prices fell dramatically, state-level regulatory schemes were introduced to stabilize production. But as one analyst notes, "even in the depths of the Depression, public authorities rarely assumed the right to determine production quotas for private industries."

The early New Deal, within the National Industrial Recovery Act, inaugurated a new state involvement. The NRA administrator promoted industrial collaboration to design industrywide codes. But government actions even at this juncture were defensive, attempting to rescue regulatory practices to stabilize production and prices in the face of intense intraindustry disputes. The final Oil Code did grant the petroleum administrator, Harold Ickes, power to determine monthly production quotas on the basis of recommendations from an industry advisory committee. Ickes attempted to further centralize his administrative powers in 1934 and 1935; opposition within the industry and in Congress prevented his doing so.

Like the temporary wartime petroleum planning efforts that would soon follow, the NRA mechanisms of government-led industrial organization did not lead to enduring government powers or responsibilities. As New Deal controls lapsed, Franklin Roosevelt in late 1935, over the objections of Harold Ickes, sent a message to the Independent Petroleum Association urging industrial self-regulation and cooperation in order to prevent government involvement. Thus the New Deal did not bring policy innovations to the oil industry. Rather, it built upon the administrative and regulatory mechanisms founded in earlier decades, fashioning a more sophisticated and efficient implementation of styles of intervention that had previously been experimented with. By the 1940s a fairly well-established pattern of state involvement had survived dramatic variations in prevailing policy and industrial concerns.

Congressional and executive officials responded, when they could, to the interests of both domestic and international companies. Thus Congress allowed tax deductions (a depreciation allowance) on domestic producers which could exceed actual levels of investments. International firms were also allowed special provisions to protect large amounts of foreign-generated income from U.S. taxation.[42] When conflicts emerged between domestic and international producers, however, domestic interests tended to win out, as the 1959 Mandatory Oil Import Quota illustrates. As foreign oil became more abundant and inexpensive, the 1959 quota protected domestic producers.[43]

The pattern of U.S. policy before 1973 was to accord special treatment to mineral and resource exploration and production through tax subsidies, and to ensure adequate prices through regulatory

[42]Stephen L. McDonald, "Taxation System and Market Distortions," in Robert J. Kalter and William A. Vogely, eds., *Energy Supply and Government Policy* (Ithaca: Cornell University Press, 1976), pp. 26–50; Wright, "Case of the United States," p. 169.

[43]Douglas Bohi and Milton Russell, *Limiting Oil Imports: An Economic History and Analysis* (Baltimore: Johns Hopkins University Press, 1978).

means. Investment decisions remained private, and the "unique role [of the state] in setting the rules of the game within which the energy market functions" was rarely considered.[44]

CRISIS AND THE LIMITS OF CHANGE

The emergence in America of a petroleum industry without significant direction from state bureaucrats had important implications for subsequent industry-government relations. With limited policy instruments at its disposal, and with congressionally mediated pressures to attend to, the federal executive used regulatory means to promote competition and stability within the industry. When supply became a national concern, executive officials encouraged production through favorable tax arrangements. Foreign exploration and production were facilitated by diplomatic initiatives.

Although the management of energy production and supply was left in private hands, executive officials did act to promote broader national energy goals. These goals included a national security interest, which became pressing during World War I, in stable and secure supplies of oil. The military, particularly the navy, found petroleum to be increasingly critical to its capacities.[45] Other government officials, with broader policy concerns, considered petroleum a strategic industry of importance to the general health of the economy. They understood stable supplies of energy at reasonable prices to be important for national economic growth, and the issue of economic growth frequently manifested itself as a concern for efficient markets in oil and energy.[46] Finally, the level of national dependence on foreign petroleum raised issues of vulnerability to and manipulation by foreign powers. This fundamentally political concern arose from the general geopolitics of nationhood.[47] Self-sufficiency, or at least some measure of national autonomy in critical industrial areas, remains a powerful underlying motivation of the state.[48]

Despite constrained institutions, U.S. executive officials found op-

[44]Robert J. Kalter and William A. Vogely, "Introduction," in Kalter and Vogely, *Energy Supply and Government Policy*, p. 11.

[45]Nash, *United States Oil Policy*, pp. 16–20.

[46]Craufurd D. Goodwin, "The Truman Administration: Toward a National Energy Policy," in Goodwin, ed., *Energy Policy in Perspective* (Washington, D.C.: Brookings, 1981), pp. 2–3.

[47]See Sen, *Military Origins of Industrialisation*, chap. 2.

[48]See Kenneth N. Waltz, "The Myth of National Interdependence," in Charles Kindleberger, ed., *The International Corporation* (Cambridge: MIT Press, 1970).

portunities to pursue national goals, most clearly in the government's encouragement of foreign exploration and production by American multinationals. Except briefly in wartime the limitations of the prevailing system and the constraints on government officials were rarely revealed. The system was successful for several generations in meeting broad national goals. The consistency of government goals and the apparent success of policy was indicated by John J. McCloy, an adviser to presidents from Franklin Roosevelt to John Kennedy, in 1974. "Strategists in the Government and Congress," McCloy argued, "did everything they could to encourage American oil companies to enter into this new area [the Middle East] so as to protect and supplement our own reserves which had been severely drained and, as a result of this foresight, oil did flow at reasonable prices, as I said, to the free world and it continued to do so for a long time. For a long period we had a stable period of supply and prices."[49] Despite its consistency and "success," however, the American system did not go unchallenged by government officials.

The Problem of State Ownership of Oil Production

Challenges to the prevailing government-business system came primarily at moments when public concern arose about petroleum supplies. The most important such moment occurred in World War II. Increasing reliance on oil imports and the problem of wartime energy shortages prompted government concern over resource availability and security. The seriousness of the perceived problem gave new weight to a government search for a national resource policy. Special wartime government planning institutions provided the opportunity for new forms of state involvement in the petroleum industry. The fate of those government initiatives provides insights into the possibilities for and limitations on institutional transformation at critical historical periods.

By 1941 the exigencies of war had brought the government into a dominant position in the petroleum industry. The chief challenge, in contrast to that of the previous decade, was to maximize production. Although Congress resisted authorization of new presidential powers to control petroleum production and distribution, Roosevelt resorted to executive action.[50] As a result, the federal government gradually

[49]John J. McCloy, testimony in Hearings before the Subcommittee on Multinational Corporations of the Committee on Foreign Relations, U.S. Senate, 93d Cong., 2d sess., pt. 5, January 24, 1974, p. 65.
[50]Goodwin, "Truman Administration," p. 52.

put in place controls over prices and production, the allocation of supply, and the direction of refining, transportation, and distribution.

Responsibility for coordination was established in the Petroleum Administration for War (PAW). The central bureaucratic figure was Secretary of Interior Harold Ickes, who was appointed petroleum coordinator for national defense on the eve of the American entry into the war. Ickes was instructed to gather information, coordinate voluntary industry collaboration, and recommend policy. His initial responsibilities were ambiguous, his powers modest. Ickes struggled to expand his powers, however, and eventually the PAW developed into an influential agency with direct access to the president.

The PAW was organized along functional lines. An industry committee was also created, the Petroleum Industry War Council (PIWC), to bring industry into policy deliberations. This organizational design, along with the staffing of PAW offices with knowledgeable industrial personnel, contributed to the agency's effectiveness.[51]

The fear on the part of both industry and government leaders of a natonal oil shortage came to a head in mid-1943. On July 14, Roosevelt called a cabinet meeting to plan American international petroleum strategy. At this juncture Ickes used his position to seek further expansion of government coordination and control of petroleum. Ickes had for several years been stressing the seriousness of the decline in the discovery rate of American petroleum reserves. In 1941, in a letter to Roosevelt, he had estimated that the United States had enough oil supplies for only fifteen years and that, if Germany were to gain control of the Persian Gulf, the Western hemisphere would fall behind Germany in petroleum production.[52]

Problems in increasing domestic oil production led industry leaders in the PIWC to urge higher prices in order to stimulate drilling. Government officials, however, turned their attention to the consolidation of access to foreign petroleum sources. In June 1943 Ickes wrote to Roosevelt urging the administration to move immediately to acquire and develop foreign reserves. He proposed that the government create a Petroleum Reserves Corporation for this purpose, drawing on earlier proposals for such a corporation which he had advanced in 1935 and 1940.[53]

[51]Nash, *United States Oil Policy*, p. 159.

[52]John Frey and Chandler Ide, *A History of the Petroleum Administration for War, 1941–45* (Washington, D.C.: GPO, 1946); Nash, *United States Oil Policy*, p. 160.

[53]Nash, *United States Oil Policy*, p. 172. See also David S. Painter, *Oil and the American Century: The Political Economy of U.S. Foreign Oil Policy, 1941–1954* (Baltimore: Johns Hopkins University Press, 1986).

Several circumstances made the Ickes proposal attractive. The first concern was with the seeming inadequacy of future American reserves. In addition, recognition was growing of the vastness of Middle Eastern petroleum resources, particularly in Saudi Arabia. Finally, and vital to the rationale for a government corporate initiative, U.S. officials and industry leaders were concerned that these petroleum deposits would come under British domination of the area.[54] Ickes also noted in his letter to Roosevelt that petroleum shortages in the coming year could impinge upon both the military campaign in Europe and industrial performance at home.[55] Indeed, the military strongly supported the proposal for direct government ownership of a petroleum firm. Although many in the State Department opposed the proposal, key proponents beyond Ickes and his Petroleum Administration for War were the Joint Chiefs of Staff and members of the Army-Navy Petroleum Board.[56]

At this juncture the American government gave consideration to participation in the Saudi concession.[57] Government support for direct participation in Saudi Arabia came from President Roosevelt himself, as well as from Secretary Ickes and Undersecretary of the Navy William C. Bullitt. American options were considered within the State Department's special Committee on International Petroleum Policy, chaired by Economic Adviser Herbert Feis, and it was Feis who gained Secretary of State Cordell Hull's assistance (if not direct support) for the establishment of a Petroleum Reserves Corporation. The State Department proposed that the PRC acquire option contracts on Saudi oil. Ickes, supported by the military, urged that the PRC acquire petroleum reserves directly, and do so by purchasing shares of the California Arabian Standard Oil Company. In June 1943 Roosevelt agreed to the Ickes plan.[58]

---

[54]Irvine H. Anderson, *ARAMCO, the United States and Saudi Arabia: A Study of the Dynamics of Foreign Oil Policy, 1933–1950* (Princeton: Princeton University Press, 1981), p. 42.

[55]Nash, *United States Oil Policy*, p. 172.

[56]Stephen Randall, "Harold Ickes and United States Foreign Petroleum Planning, 1939–45," *Business History Review* 57 (Fall 1983), 373; see also Stephen J. Randall, *United States Foreign Oil Policy, 1919–1948: For Profits and Security* (Kingston: McGill-Queen's University Press, 1985).

[57]Anderson, *ARAMCO*, p. 42.

[58]The board of directors of the PRC was to consist of the secretaries of state, interior, war, and the navy. The State Department was to have veto rights over PRC activity, retaining the responsibility to conduct negotiations with foreign governments. Although the initial proposal for the PRC came from the State Department, the wartime concern over declining domestic petroleum reserves and the navy's assertive support allowed the bolder Ickes plan to be adopted. See Randall, "Harold Ickes and U.S. Foreign Petroleum Planning."

Secret negotiations were begun to buy shares in CASOC. After difficult talks, CASOC eventually proposed that the government should purchase a one-third interest in the company. In addition, the U.S. government would have a right to buy up to 51 percent of CASOC production during peacetime and 100 percent during war, and it could block the sale of company oil to third parties. Apart from these restrictions, CASOC would proceed under ordinary commercial rules.

Most accounts of the negotiations, and the ultimate failure of government initiatives, have centered on opposition to the plan from CASOC itself.[59] However, evidence indicates that opposition from other companies, namely Standard Oil of New Jersey and Socony-Vacuum, was decisive.[60] Industry opposition was centered in the Foreign Operations Committee of the PAW. This wartime association of petroleum executives favored the private development of Saudi reserves, with American diplomatic support. Irvine Anderson notes that "this was formidable opposition, coming as it did from a bloc of companies that up to that point had contributed mightily to Ickes' success as Petroleum Administrator for War. Ickes' rather formidable power within the bureaucracy rested heavily on his close relationship with Roosevelt, which in turn was based on his proven ability to get results. If this corporate revolt were to undermine cooperation with the PAW itself, the consequences would be serious."[61]

Although the formal reason for the withdrawal of the government purchase arrangement was inability to agree on terms, opposition within the loose affiliation of petroleum executives was crucial to the outcome. Yet a complete explanation for the demise of the Petroleum Reserves Corporation must go beyond business opposition. Opposi-

---

[59]Raymond F. Mikesell and Hollis B. Chenery, *Arabian Oil: America's Stake in the Middle East* (Chapel Hill: University of North Carolina Press, 1949), pp. 91–92; Nash, *United States Oil Policy*, p. 173; George Stocking, *Middle East Oil* (Nashville: Vanderbilt University Press, 1970), pp. 98–99; Mira Wilkins, *The Maturing of Multinational Enterprise: American Business Abroad from 1914 to 1970* (Cambridge: Harvard University Press, 1974), pp. 277–78; Sampson, *Seven Sisters*, pp. 96–97; Robert B. Kruger, *The United States and International Oil: A Report for the Federal Energy Administration on U.S. Firms and Government Policy* (New York: Praeger, 1975), pp. 50–51; Krasner, *Defending the National Interest*, pp. 190–97; Michael B. Stoff, *Oil, War, and American Security: The Search for a National Policy on Foreign Oil, 1941–1947* (New Haven: Yale University Press, 1980), pp. 84–86; and David Aaron Miller, *Search for Security: Saudi Arabian Oil and American Foreign Policy* (Chapel Hill: University of North Carolina Press, 1980), pp. 81–82.

[60]Anderson, *ARAMCO*, p. 56. See also Robert O. Keohane, "State Power and Industry Influence: American Foreign Oil Policy in the 1940s," *International Organization* 36 (Winter 1982), 166–83.

[61]Anderson, *ARAMCO*, p. 63.

tion within the administration, notably from the State Department, made the PRC problematic even within the context of the goals and interests of American foreign policy. Also, several investigatory committees—the Special Committee Investigating the National Defense Program and the Special Committee Investigating Petroleum Reserves—were skeptical about specific Ickes plans for government petroleum projects in Saudi Arabia. What becomes evident is that an array of groups and agencies served to dilute Ickes's position within government and the legitimacy of his proposals.

The failure of the Ickes initiatives takes on substantial significance from the vantage point of later decades—a significance not lost on officials and politicians in the midst of the oil shocks of the 1970s. In hearings on multinational corporations and American foreign policy, Senator Frank Church made this observation:

> Whether this idea [for a national oil company] was a good or bad idea is debatable, but the proposal for an independent U.S. Government capability to formulate a national petroleum policy, which was the objective of the Ickes' proposal, regardless of its form, never bore fruition. It was blocked by industry pressures. Thus was woven the basic pattern which has, since that time, formed the basis of Government-industry relations—the Government to provide the diplomatic and financial support for the industry in its operations abroad, but denuded of any institutional capability to formulate a policy of its own or to oversee the operations of the American petroleum industry abroad.[62]

Opposition both within the government and among leaders in the petroleum industry prevented the Ickes plan from inaugurating a new government role in oil and energy. Not until the 1970s would such proposals again find their way into the mainstream of the policy process. And even in the midst of the energy crisis of the 1970s, such proposals were wielded with much less conviction. In a study of government-industry relations in the petroleum area one analyst correctly concludes that

> the most significant consequences of this series of unsuccessful forays into international petroleum affairs was that the U.S. Government thereafter implicitly left the function of control and supervision over the international petroleum system to the multinational petroleum companies. Unlike the British, French and Dutch, with their governmental interests in BP, CFP and Shell, respectively, the U.S. Government now,

[62]Hearings before the Subcommittee on Multinational Corporations of the Committee on Foreign Relations, U.S. Senate, 93d Cong., 2d sess., pt. 4, January 30, 1974, p. 2.

73

for the most part, divorced itself from the international petroleum industry. These events signified as well a virtual cessation of the Government's efforts to obtain an information base independent of the companies, which might help it to formulate petroleum policy and take independent action.[63]

The 1943 episode was a watershed in industry-government relations. At that historical moment a departure was contemplated at the highest levels of government from long-established patterns of autonomous public and private spheres. The government's impressive wartime controls over domestic production and distribution were possible, however, only in the context of a national crisis. The administration of government control was effective primarily because the industry itself was brought into the formation and implementation of policy. Nonetheless, an aggressive bureaucratic leader was able to assemble an agency to advance extraordinary proposals to bring the government directly into the industry. The fate of his agency reveals the limits of that process, even in times most favorably disposed to institutional innovation.

Finally, the Ickes experience reveals another aspect of American state structure. The PAW and its planning pretensions were not sufficiently established to survive the wartime emergency. The PAW, as one study notes, "functioned effectively . . . and . . . even manag[ed] to self-destruct when the war ended."[64] The institutional growth and decline of the PAW, where little wartime government capacity survived the war itself, signals a broader administrative pattern.

Energy Planning and Administration

Historically, American planning capacities instituted or augmented during times of war or crisis have been dismantled or radically cut back with the return to normality. This pattern is clear in the energy resource area, with its cycles of state concern and complacency over energy supply and security. At various junctures of perceived shortages or threat, and in particular during World War II, executive officials have taken steps to develop resource planning capacities. These halting and ultimately unsuccessful efforts to develop comprehensive capacities within government left a fragmentary, ad hoc system of administrative responsibility for energy.

[63]Krueger, United States and International Oil, pp. 51–52.
[64]Aaron Wildavsky and Ellen Tenebaum, The Politics of Mistrust: Estimating American Oil and Gas Resources (Beverly Hills, Calif.: Sage, 1981), p. 98.

The most ambitious federal organization for resource planning, which dealt with energy issues but also ranged widely on issues of economic and industrial order, was the National Resources Planning Board. From 1939 to 1943 the NRPB consolidated and attempted to coordinate a variety of planning functions within the federal government. The agency had its roots in earlier planning bodies that had sprung from the early emergency programs of the New Deal. Its first incarnation was as the National Planning Board. Sympathetic to proposals for a national planning organization, Harold Ickes incorporated the National Planning Board into the Public Works Administration in 1933. This planning organization survived several more transformations to become the focal point for public policy planning, coordination, and research within the Roosevelt administration. The board received independent status from the Reorganization Act of 1939, which situated it in the Executive Office of the President, formalizing the operations of the staff. Yet the NRPB still rested precariously within the executive establishment. The agency was not well received in Congress, and indeed it was established and maintained in its early years out of discretionary emergency funds held by the administration.[65]

Composed of a small staff of professional economists and planners, the NRPB had as its central mission to produce reports on manpower, social, and resource policy. It operated as an island of analytic expertise unrelated to the functional responsibilities of executive departments. Although the NRPB attempted to cooperate with established departments, its influence was manifest primarily in the reports it produced and in periodic meetings of its staff with the president.[66] In this body evolved the rationale and instruments of an interventionist, Keynesian, postwar economic program of full employment. The board also established itself as the most important federal agency involved in gathering information on and in analysis of energy and natural patterns.[67]

In 1938 the National Resources Committee, a special planning group under the direction of Secretary Ickes, prepared a comprehensive study entitled *Energy Resources and National Policy*, which marked the first step toward comprehensive energy planning within the government.[68] The report provided the rationale for the activities of the

[65]Barry D. Karl, *Charles E. Merriam and the Study of Politics* (Chicago: University of Chicago Press, 1974), pp. 270–72.

[66]Marion Clawson, *New Deal Planning: The National Resources Planning Board* (Baltimore: Johns Hopkins University Press, 1981), p. 9.

[67]Ibid., pp. 107–24; Charles E. Merriam, "The National Resources Planning Board," *Public Administration Review* 1 (Winter 1941), 116–21.

[68]Clawson, *New Deal Planning*, p. 122.

council. "It is time now to take a larger view," the report argued, "to recognize more fully than has been possible in the past that each of these energy resources affects the others, and that the diversity of problems affecting them and their interlocking relationships require the careful weighing of conflicting interests and points of view."[69] In energy resources the NRPB provided a setting for the gathering and organization of expertise that otherwise was widely scattered. With the demise of the NRPB, that professional expertise would again be fragmented in disparate public and private organizations.

The NRPB had recruited an impressive group of economists and policy analysts and articulated a strong appeal for planning and coordination. It failed, however, to establish itself firmly in the permanent federal bureaucracy. The proximate cause of its demise was termination by Congress of its funding. But the more important reasons lay in its claims to policy responsibility, which competed with those of a functional and clientilistic federal bureaucracy that was already well entrenched. Consequently, toward the end of World War II the NRPB was divested of its responsibilities, which came to reside in fragmented form throughout the federal establishment. Economic reports on business and unemployment trends, for example, were placed with the Council of Economic Advisers in 1946, and science policy found its way into the National Science Foundation in 1950. As far as energy resources were concerned, production and conservation programs dealing with oil and coal were given to the Department of Interior.[70] The NRPB had responded to Roosevelt's long-standing interest in comprehensive natural resource planning, but it could not gain an institutional foothold to survive even the end of the Roosevelt administration.[71]

The devolution and dispersion of planning went even further than this. The failure of federal administrative planning and coordinative responsibilities left these activities to organizations outside the established structure. In energy and resource planning two developments were important. Long-range analysis of resource security and supply was handled through temporary commissions. The most important example was the president's Materials Policy Commission (the Paley Commission), which in the mid-1950s provided a systematic study and appeal for federal planning initiatives but failed to spur signifi-

[69]*Energy Resources and National Policy,* Report of the Energy Resources Committee to the Natural Resources Committee (Washington, D.C.: GPO, 1939), p. 1.

[70]Clawson, *New Deal Planning,* pp. 245–46.

[71]Arthur M. Schlesinger, Jr., *The Age of Roosevelt—The Coming of the New Deal* (Boston: Houghton Mifflin, 1959), p. 350.

cant government action. The other development, following from the eclipse of the NRPB and a strong federal planning presence, was the growth of resource policy institutions and programs at private foundations. In fact the nonprofit institution Resources for the Future was a direct descendant of the NRPB and the Paley Commission.[72]

The Paley Commission report, a landmark study of national energy problems, was a response to general concern over the depletion of domestic energy fuels. For over a year staff experts gathered from various government departments, business, and several universities. The assembled group of energy specialists, economists, and policy analysts gave the Paley Commission a unique capacity to develop broad, national, and long-term perspectives on energy problems and policy. Such a sustained gathering of professionals, resembling the expert resources brought together in the NRPB, would not be seen again until the 1970s. Before and after the commission, however, energy policy resided primarily in the Interior Department and in several remote congressional committees.

The Paley Commission concluded that demand for energy in the industrial world would vastly increase, doubling by the mid-1970s. Conventional sources of energy would fall short in the next few decades, it argued, requiring the introduction of new energy sources and technologies. While recognizing the unfolding of powerful historic forces, the commission did not recommend major new government powers or responsibilities. Precisely because it cast its analysis in terms of world energy markets and Western security interests, however, it did strongly urge a comprehensive energy policy: "The Commission is strongly of the opinion that the Nation's energy problem must be viewed in its entirety and not as a loose collection of independent pieces involving different sources and forms of energy."[73] The commission noted that "this implies no increase in Government activity; it well might mean less. It does mean that the multiple departments, bureaus, agencies and commissions which deal with separate energy problems must be less compartmentalized."[74]

The commission delivered its recommendations to President Truman in June 1952. Truman in turn gave the National Security Resources Board responsibility for evaluating the commission's findings and making specific proposals. The NSRB and various govern-

[72]Clawson, *New Deal Planning*, pp. 250–52; Milton Russell, "Energy Politics Looking Back," in Martin Greenberger, *Caught Unawares: The Energy Decade in Retrospect* (Cambridge, Mass.: Ballinger, 1983), pp. 29–66.
[73]President's Materials Policy Commission, *Resources for Freedom* (Washington, D.C.: GPO, 1952), 1:129.
[74]Ibid., 1:129.

ment agencies greeted many of the commission findings with approval. Disputes broke out over whether government should collect its own energy data or leave that task to the industries themselves. The Interior Department proposed that the commission's objectives could best be accomplished through an expansion of responsibilities within that department.[75] With the impending departure of the Truman administration, however, substantive organizational reform was not accomplished. The expert professional capacities of the Paley Commission scattered back to separate public and private organizations.

As this episode suggests, key U.S. executive officials and experts have called at important junctures for the expansion and centralization of energy policy and planning. The most important attempts to establish a high-level, coordinated planning body have come at moments of public concern over energy supply and security. However, proposals such as those by the Paley Commission and organizational experiments such as the NRPB failed to gain institutional footholds.

CONCLUSION

The organizational structures of the state are explicable in terms of larger historical processes. The crucial explanation for both government-business relations and state administration stems from the *sequence* of historical processes. The rise of large-scale business before that of large-scale government bureaucracy in the United States explains important aspects of subsequent relationships between the two. Also, and in contrast to the European experience, the establishment of representative institutions in the United States preceded the establishment of national administrative institutions.

Although moments of crisis periodically generated proposals and institutional experimentation aimed at developing new state capacities, these changes did not become a permanent part of the organizational structure of national government. Existing institutional relations of business and government preserved a minimalist role for the state. National energy goals were pursued within a fragmentary organizational context that conferred little unitary power on policy makers and provided few instruments for shaping private behavior.

Moments of crisis provided apportunities for executive officials to challenge this institutional regime. Such moments occurred primarily

---

[75]Goodwin, "Truman Administration," pp. 60–61.

during the two world wars and the years preceding the Korean War. However, obdurate congressional resistance to planning, the weak bureaucratic position of wartime planning organizations, and the fleeting perception of crisis conspired to debilitate and diffuse proposals for institutional change. Consequently, institutional patterns in the energy area before 1973 exhibited a striking continuity. There was no centralized administrative body charged with programmatic policy planning, and the instruments of national policy remained limited and indirect. These institutional circumstances provided the foundation for political maneuvering and policy change as state officials confronted demands for energy adjustment in the 1970s.

CHAPTER FOUR

The Limits of International Cooperation

> In the short term, it is possible to conceive any number of bilateral deals that can be made between major consuming countries and major producing countries. In the short term, it is possible to see how particular producing countries can enrich themselves by an unrestrained use of their temporary strong bargaining position. But in the long term, it is bound to lead to disaster for everybody. It is particularly a case where the common interest is also everybody's selfish interest.
>
> Secretary of State Kissinger, January 3, 1974

The first line of action for American executive officials in the wake of the oil price shocks of 1973–74 was diplomatic. Led by the State Department, administration officials sought to draw together the industrial consuming nations in order to confront OPEC and negotiate the rollback or stabilization of oil prices. Despite the persistence of American efforts, these cooperative schemes did not succeed. The International Energy Agency (IEA), an international organization of modest scope, was created, and it provided stand-by arrangements for emergency oil sharing. But it was unable to play the role American officials had envisaged for it. Programmatic cooperative agreements also failed to take hold during the second oil shock. Government leaders did make pledges in 1979 to reduce oil import levels, but these collective statements of intention largely proved unavailing.

Despite the magnitude of energy adjustment problems, and despite what appeared to be a basic mutuality of interests among the industrial consuming nations, policy coordination remained elusive. This chapter seeks to account for the persistence of American efforts to achieve a common solution to those problems and for the failure of those efforts.

In Chapter 2, I argued that international cooperative agreements (international offensive and defensive adjustment) are part of a larger set of adjustment strategies and that the attractiveness of each option is related to the state's domestic capacities for adjustment and the goals that arise from the state's international position. Applied to the case at hand, these considerations suggest that U.S. officials favored the cooperative adjustment strategy for two sorts of reasons. First, government leaders were constrained in their ability to impose the costs of adjustment on the economy, primarily by domestic price controls on petroleum. These controls, enacted prior to 1973–74, shielded consumers from the full force of external price increases and gave price windfalls to some segments of the petroleum industry. In response to these interests, Congress preserved and extended price and allocation controls in the years that followed. Swift market-driven adjustment to escalating petroleum prices remained beyond the reach of executive officials.

Second, the privileged international position of the United States uniquely favored international strategies of adjustment. Several features of the American international position were important. By virtue of its large absolute share of international oil imports, which satisfied a small relative share of domestic energy consumption, the United States was favorably positioned to lead Western importing nations in a response to OPEC. Its dominant international political position also gave the United States a broad set of objectives in the oil price revolution. Accepting higher petroleum prices in the West meant accepting slower growth, greater export competition, and bilateralism, and encouraging demands by developing countries for a New International Economic Order—all of which ran counter to American postwar ideals. The United States, in sum, was interested not only in how it adjusted domestically but in how other advanced industrial states adjusted as well. Together, domestic constraints on energy adjustment and a commanding international position led to an initial American emphasis on international strategies of adjustment.

The failure of U.S. cooperative proposals was rooted in the differences among importing nations in the costs and objectives that attended energy adjustment. Whereas the United States found it particularly difficult to pass higher oil prices through its domestic economy, other industrial states accepted those higher costs. Consequently, the United States put the pricing issue at the top of its foreign policy agenda, whereas its industrial partners were more concerned with access to supply and favorable diplomatic relations with the Arab states. By adjusting to higher prices and pursuing a different set of objectives, these countries in effect exited from multilateral adjustment schemes, and what followed was the failure of the American grand design.

OPPORTUNITIES FOR COOPERATION

Prior to the 1973–74 oil embargo, the industrial nations had not attempted formally to coordinate energy policies. The oil crisis of 1956–57, for example, had been met by unilateral American action. In response to conflict with Britain, France, and the United States, Egypt nationalized the Suez Canal in July 1956. This action led to an international crisis that culminated in the invasion of Egypt by Israeli, French, and British forces. Although the invasion was cut short by U.S. pressure, the Suez Canal was blocked, cutting the major transportation route of oil from the Persian Gulf to Europe. The American government, although faced with initial resistance from the Texas Railroad Commission, was able to increase domestic production and, with cooperation from the international oil companies, ease European shortages during the winter of 1956–57. Because the United States had access to large, readily available domestic reserves, it could compensate for international dislocations and stabilize supplies.[1] By the early 1970s, however, the American position in the world petroleum market had changed. Domestic production had peaked, and the United States had begun to import larger amounts of foreign oil. In 1973 the U.S. government was unable to match the shortfall in OPEC production with expansion of its own.

The absence of a power capable of countering OPEC restrictions made cooperation among oil-consuming nations more vital if Western governments were to control supplies and prices of international petroleum. A variety of opportunities for cooperation existed to lower or redistribute the costs of oil shortages and higher prices. Both petroleum crises in the 1970s began with the curtailment of a portion of OPEC production. In 1973–74 the attempt by Arab OPEC member countries to embargo the United States created a shortfall of about 9.8 percent of available world supplies.[2] The scramble for crude oil by consuming nations, each concerned to maintain its access to imported oil, pushed prices on the spot market dramatically higher and gave

[1]See Robert O. Keohane, *After Hegemony* (Princeton: Princeton University Press, 1985), pp. 169–174.
[2]Noncommunist production fell from 47.8 billion barrels a day to 43.2 million barrels a day between September and November 1973. Philip K. Verleger, Jr., *Oil Markets in Turmoil: An Economic Analysis* (Cambridge, Mass.: Ballinger, 1982), p. 33.

OPEC additional opportunities to raise its prices. In 1979 the tightening of the world oil market started when Iran curtailed oil production in the wake of its revolution. The shortfall in late 1978 was approximately 6.3 percent of world supply.[3] This tightening of supply was, as in the first crisis, followed by the bidding up of prices on the spot market and a rise in official OPEC prices. Competitive bidding by consumer nations on the spot market was even more consequential for OPEC pricing in the second crisis, because by the late 1970s more international petroleum transactions were being conducted through the market and less through long-term contract.[4] In both cases the supply shortfall was small, but the competitive actions it unleashed and the pricing actions it allowed disadvantaged all consuming nations. These supply and pricing problems thus provided substantial opportunities for mutual gains through cooperation.

The most immediate form of cooperation would have entailed agreement by the consumer nations to restrain import demand. By abstaining from unilateral efforts to increase supplies, or by going further and actually reducing current import levels, consumers could reduce pressures on already limited world supplies. In the absence of such an agreement, and with each nation pursuing its own short-term objective of gaining access to existing supplies, prices would rise higher than when nations agreed to moderate demand.

Cooperation could also take the form of longer-term efforts by consumer nations to conserve energy and generate alternative sources of supply. Such efforts would also alter the supply and demand pattern and, by so doing, undercut the pricing power of the cartel. Whereas the cooperative actions noted above involve immediate steps to limit oil imports, the plan here involves more gradual efforts to address the underlying structure of demand and supply. Collective agreement would increase the efficacy of individual actions by governments.

In both cases—whether demand restraint or conservation and production initiatives—consuming nations could pursue actions individually. Yet unilateral actions would have less overall impact on prices and supplies. Indeed, without agreement each government was likely to pursue separate policies of national gain even as those actions left all nations worse off than they would have been if a collective agreement had been achieved. In this sense the problems of agree-

---

[3]Between October and November noncommunist world production fell from 62.9 million barrels a day to 58.9 million barrels a day. Ibid., p. 33.

[4]See Thomas Neff, "The Changing World Oil Market," in David A. Deese and Joseph Nye, eds., *Energy and Security* (Cambridge, Mass.: Ballinger, 1981).

ment by consuming nations over energy adjustment were driven by the dilemma of collective action.[5]

These forms of cooperation can be characterized as international offensive strategies of adjustment. The objective is to engage in cooperative actions that serve to counter or mitigate unwanted international economic change by going to the source of that change. In this case, the strategies involved joint actions by consuming nations to roll back or moderate oil price increases or to engage in indirect bargains (between consuming nations and between those nations and OPEC) to stabilize oil prices and supply. Consuming nations could also take internationally defensive cooperative actions: emergency oil-sharing agreements that spread the burden of reduced oil supplies among industrial importing nations, and international financial agreements that cushion balance of payments problems generated by the higher costs of petroleum imports.[6] Such actions do not go to the source of the international economic disturbance; they aim, rather, at collectively equipping the consumer nations with a means to cope with the most onerous consequences of the disturbance.

The United States advanced cooperative proposals along both international offensive and defensive lines after 1973. Not only did other consuming nations have strategies that addressed more immediate political and security objectives, however, but the United States itself was unable to generate the domestic changes necessary to make the multilateral cooperative strategy successful.

AMERICAN INTERNATIONAL ENERGY STRATEGY

The U.S. government was the most persistent and forceful advocate of a cooperative response by industrial consumer nations to rising oil prices in the early 1970s. This strategy, articulated by the State Department and Secretary of State Henry Kissinger, initially sought to salvage the world petroleum order by confronting OPEC directly.

---

[5]The seminal presentation of the dilemma of collective action is Mancur Olson, *The Logic of Collective Action* (Cambridge: Harvard University Press, 1965). The extension of this logic to alliance cooperation is found in Olson and Richard Zeckhauser, "An Economic Theory of Alliances," *Review of Economics and Statistics* 48 (August 1966), 266–79. Robert Keohane also presents the difficulties of cooperation by consuming nations after 1973 as a dilemma of collective action. See *After Hegemony*, pp. 223–37.

[6]For a discussion of the full range of cooperative issues as noted by an observer at the time, see Walter L. Levy, "World Oil Cooperation or International Chaos," *Foreign Affairs* 52 (July 1974), 690–713.

American efforts to counter OPEC directly were blunted by resistance from allies and intractable political constraints at home. Government officials sought to adapt their international strategy after its initial failure, but in due course they grudgingly abandoned it in favor of a sequence of domestic adjustment strategies.

American officials showed an interest in an international solution to energy shortages before October 1973. Urging a united front against OPEC, one State Department official wrote in early 1973 that the Western importing nations should create an "international authority" that would "avoid cutthroat competition" for available energy supplies. He argued that "such competition could drive prices far higher than we can presently imagine."[7] In April 1973 the State Department disclosed an ongoing planning effort to establish an international organization of oil-importing countries whose primary purposes would be to allocate petroleum supplies in times of shortage and to develop practices to avoid an "international price war." Although State Department officials differed about the possibilities for a "buyers' cartel," most agreed that cooperative measures would improve the joint bargaining position of industrial nations with OPEC.[8] In responding to the proposal, the French and Japanese governments foreshadowed the positions they would take in the months ahead. France acknowledged the need in principle for some cooperation but argued that European cooperation should come first. The Japanese insisted that producers not be provoked unnecessarily.[9]

The logic of the American position on the unfolding energy problem was revealed in comments by the State Department's undersecretary for economic affairs, William J. Casey. In a discussion of the energy predicament in June 1973, Casey acknowledged that rising American consumption of imported oil was aggravating the situation by increasing world demand for Middle Eastern oil and thereby putting upward pressure on prices. But the U.S. government, he maintained, favored international responses to the problem and, therefore, urged consultation and cooperation in areas such as energy conservation, production, and technology. In effect, Casey argued that although the United States was contributing to the problem, the

---

[7] James E. Akins, "The Oil Crisis: This Time the Wolf Is Here," *Foreign Affairs* 51 (April 1973), 486.

[8] See Edward Cowan, "US Plans World Group of Oil-Importing Nations," *New York Times*, April 16, 1973, p. 28.

[9] On France see *New York Times*, May 24, 1973, p. 75; on Japan, Alexander K. Young, "Energy: View from Tokyo," *New York Times*, July 22, 1973, p. 3:6.

solution would need to be arrived at jointly among industrial consuming nations.[10]

Confronted by the October embargo, American officials redoubled their efforts to fashion a united response to OPEC. They emphasized the virtues of multilateral rather than bilateral or regional responses to problems of oil supply and pricing, and they did so for both political and economic reasons. Politically, the movement toward deals by individual countries with oil-producing nations, according to State Department officials, threatened to fragment the liberal, multilateral trading system into rival blocks. Addressing his remarks to Europeans, Kissinger argued in December 1973 that unless cooperative measures were taken to alleviate the crisis, the industrial world would be faced "with a vicious cycle of competition, autarchy, rivalry and depression as led to the collapse of the world order in the thirties."[11] The United States, therefore, found itself opposing actions by other governments, particularly the French and the Japanese, to forge bilateral deals with OPEC producers. Bilateral oil trading relationships threatened the postwar norms of nondiscriminatory trade and open markets.

The United States agrued, moreover, that the prevailing higher oil prices were neither inevitable or necessary. If consumer nations cooperated to limit import demand and stimulate their own production, they could undercut OPEC pricing, and more moderate prices would prevail. This remained the core of American international energy policy throughout the Kissinger period. The American objective, as a senior State Department official noted in congressional testimony, was to "bring about a basic shift in the supply/demand balance in the world oil market. This will reduce our vulnerability to foreign supply disruptions, reduce the ability of a small group of countries to manipulate world oil prices arbitrarily and enable prices to approach their long-term equilbrium level."[12] Only through collective action by consumer countries could competitive bidding be avoided and oil prices be moderated or reduced.

The American opposition to a fragmented consumer response to

[10]William D. Smith, "Casey Urges World Group of Oil Users," *New York Times*, June 22, 1973, p. 43.

[11]Quoted in Dankwart A. Rustow and John F. Mugno, *OPEC: Success and Prospects* (New York: New York University Press, 1976), p. 51. See also Steven A. Schneider, *The Oil Price Revolution* (Baltimore: Johns Hopkins University Press, 1983), p. 257.

[12]Charles Robinson, Undersecretary for Economic Affairs, Department of State, "U.S. International Energy Policy," Hearings before the Subcommittee on International Resources, Food, and Energy, Committee on International Relations, U.S. House of Representatives, 94th Cong., 1st sess., May 1, 1975, p. 6

OPEC, exemplified in the pursuit of bilateral negotiations, was thus both political and economic. The connection between these arguments was articulated by the American secretary of state in February 1974: "What we believe is going to be disastrous for the world economy is if bilateral deals are made unconstrained by any general rules of conduct, because we believe that this will either stabilize prices at too high a level or bid prices up even higher and in general create a relationship among the major consuming nations of economic warfare—which inevitably will affect, in time, their political relationship."[13] Maintaining the integrity of a multilateral approach to international economic problems, preserving political relations among Western industrial nations, and developing a coordinated program to counter OPEC went hand-in-hand.

The most public call by the United States for a united, multilateral response came in December 1973 with Secretary of State Kissinger's speech to the Pilgrims of Great Britain in London. Kissinger argued that the crisis of supply was a long-term problem that would require collaborative programs by consuming nations to develop new methods of conservation and sources of supply. The secretary of state called for the formation of an Energy Action Group, composed of leading public officials and other representatives from Europe, the United States, and Japan. This group's mandate would be to "define broad principles of cooperation, and it would initiate action in specific areas" such as stimulating conservation and production of energy and joint programs in research and development.[14] Kissinger's declaration made the goal of consumer unity a central element of American foreign policy.

The Washington Energy Conference

To put the American international plan into action, the Nixon administration in January 1974 called for a conference of the industrial consuming nations to meet in Washington the following month. Again Secretary Kissinger, in making the announcement, put forward the arguments for cooperation. Importing nations, he argued, had a common interest in restraining competition for supplies, in collaborating to generate additional sources of energy, and in bargaining with the producing nations not only on price but also on the

[13] "Secretary Kissinger Discusses Washington Energy Conference," *News Release*, Bureau of Public Affairs, Department of State, February 13, 1974.

[14] "Secretary Kissinger Calls for International Energy Cooperation," Department of State, *Selected Documents* no. 3 (Washington, D.C., December 1975), p. 4.

trade and monetary distortions that price increases were creating. In the absence of such cooperation, Kissinger warned of a severe world depression. Such an "economic disaster" would, he argued, be driven by unrestrained competition over limited supplies, the result of industrial nations attempting to solve their problems individually.[15]

The central challenge Kissinger and other administrative officials confronted in early 1974 was the movement by other importing nations toward individual deals with producing nations to protect their supplies of oil. American efforts to forestall these developments were renewed after the French government announced it had completed the first phase of a bilateral deal with Saudi Arabia for long-term oil supply. While not directly criticizing that deal, Secretary Kissinger emphasized the American view that "unrestricted bilateral competition" would be "ruinous for all countries concerned." He also introduced an implicit warning of retaliation, noting that the United States might seek its own arrangements to protect energy supplies. It was an option, he emphasized, the United States wanted to resist taking.[16]

The Europeans and Japanese were not eager to endorse the American plan of action if consumer cooperation was to serve as a "counter-cartel" and thus antagonize the OPEC countries. The French objected most vigorously to the proposed Washington Energy Conference, choosing to pursue instead its own strategy of a European-Arab dialogue. Japan and most European governments maintained a middle position between France and the United States. Before the Washington conference the foreign ministers of the European Community adopted a joint position. Following the French lead, they held that the Washington conference was not to be used to confront the oil-producing countries, that no new consumer nation organization should be established, that European nations must remain free to establish direct relationships with OPEC producers, and that a formal dialogue between developed and developing countries should be pushed forward.[17]

Government representatives at the Washington Energy Conference were able to agree upon a modest, collective course of action, its centerpiece an emergency oil-sharing plan. But agreement, modest as

---

[15]R. W. Apple, "Nixon Is Planning Appeal on Energy to 20 Countries," *New York Times,* January 4, 1974, p. 1.

[16]"Global Aspects of Energy Crisis Discussed by Secretary Kissinger, William Simon," *News Release,* Bureau of Public Affairs, Department of State, January 10, 1974; see also "Kissinger Bids Nations Unite in Crisis," *New York Times,* January 11, 1974, p. 10, and Henry Kissinger, *Years of Upheaval* (Boston: Little, Brown, 1982), p. 902.

[17]See Robert Lieber, *Oil and the Middle East War,* Harvard Studies in International Affairs no. 35 (1976), p. 21.

it was, was achieved only with substantial American concessions. The United States conceded, for the first time, that its domestic oil supplies would be included in the emergency sharing plan.[18] American officials also agreed to delete references to the prohibition of direct deals between consuming and producing nations from the final resolution. The U.S. delegation had proposed a code of conduct to govern bilateral agreements with OPEC producers, but this proposal was eventually dropped.[19] Most important, the United States abandoned its original view that an organization of consumer nations could be formed to articulate a common bargaining position and, by so doing, undercut the oil cartel.

To get even this limited agreement the Nixon administration was forced to pressure the European governments to eschew the French approach. American officials warned that if the conference failed to reach agreement, U.S. markets and the security commitment to Europe could not be assured. Addressing the Washington conference, President Richard Nixon argued that "security and economic considerations are inevitably linked and energy cannot be separated from either." Secretary of State Kissinger told German foreign minister Walter Scheel that the United States would reconsider its troop commitment to the Federal Republic unless the Europeans supported the establishment of an Energy Coordinating Group and rejected the French position.[20]

The Washington conference produced agreements in four areas. First, the countries agreed on joint action to allocate oil in times of emergency and to pursue energy conservation and R & D. Second, financial and monetary measures were to be developed to avoid competitive currency depreciation and to strengthen international credit facilities. Third, an Energy Coordinating Group was established to implement the agreement. Finally, the countries agreed to engage in a dialogue with the oil-producing nations. In November 1974 the sixteen-member International Energy Agency was established.

Outwardly the Washington conference appeared to be a victory for the American multilateral strategy, yet the success was more one of form than of substance. The French refused to participate in the Energy Coordinating Group and did not join the IEA. Nor did the February conference diminish the movement toward bilateral and regional agreements between European and producing countries. In-

[18]See discussion in Rustow and Mugno, *OPEC: Success and Prospects*, p. 59.
[19]Schneider, *Oil Price Revolution*, p. 260.
[20]Nixon is quoted in *New York Times*, February 15, 1974, p. 43; for Kissinger see Lieber, *Oil and the Middle East War*, p. 49.

deed, the following month the EEC Council of Ministers agreed to a program of economic, technical, and cultural cooperation with twenty Arab countries.[21] The French strategy of a broad European-Arab accord still had support within the European Community. The governments of Europe, particularly the British and the Germans, continued to pursue a middle course. They were unwilling to risk a breakdown of relations with the United States and therefore agreed to participate in the limited American program of multilateral cooperation. But they were also unwilling to abandon the opportunities for bilateral and regional agreements with Arab oil producers led by the French.[22] The result was a failure of the American strategy for a unified response by consumer countries to OPEC actions on supply and pricing. The French continued to insist on both the need for a dialogue with the producing nations and the legitimacy of separate oil agreements. The United States, though moderating its objectives to gain agreement on the IEA, continued through 1974 to search for a unified way to challenge OPEC.

EUROPEAN AND JAPANESE BILATERALISM

The central threat to Kissinger's multilateral strategy was that industrial consuming nations might pursue bilateral deals with OPEC producers, thereby undercutting cooperative efforts at demand restraint and collective negotiations over price stability. As importing nations scrambled for separate deals with producing nations, they competed for unilateral advantage, strengthening the position of producers and leaving the consuming nations as a whole less well off.

Although the pursuit of national advantage was implicit in the French bilateral initiatives, Paris based its opposition to the American multilateral proposals on broader considerations, including a difference in view concerning the nature of the international petroleum problem. Whereas the United States thought world prices could be lowered, France believed that oil prices were likely to continue to rise and therefore sought to contract for supplies at prevailing prices.[23] Moreover, the French government attempted to strengthen its position throughout the developing world by its conciliatory policy toward OPEC. French president Georges Pompidou was also able to strength-

[21]Schneider, *Oil Price Revolution*, p. 261.
[22]Lieber, *Oil and the Middle East War*, p. 23.
[23]Schneider, *Oil Price Revolution*, p. 259.

en his domestic political position by pursuing an antagonistic policy toward the United States.

Beyond these broader political considerations which informed the French position, bilateral deals were attractive to consuming nations because they secured sources of supply and, when part of a larger trade package, helped pay for the imports. In the immediate aftermath of the October embargo, for example, the French negotiated a three-year contract with the Saudi Arabian government at a price below the official OPEC price. At the same time the French government began negotiations with the Saudis for a twenty-year contract that would guarantee access to enough oil to cover roughly a quarter of French consumption. Through its state-owned company, Elf-Erap, the French also negotiated deals with Libya and Iran.[24]

Japan also pursued bilateral deals, seeking not only to gain secure sources of supply but also to promote larger economic projects with oil-producing nations. In early 1974 the Japanese agreed to lend Iraq one billion dollars and to initiate refining, petrochemical, and other industrial projects in exchange for a ten-year contract for crude oil and petroleum products. The Japanese, with an export economy dependent on the maintenance of an open, multilateral system, were more hesitant than the French to compete for bilateral deals with Middle Eastern oil producers. Also, their oil supplies from the Persian Gulf were protected by the United States. Consequently, Japanese policy preserved bilateral access to oil producers but also moderated competition with other industrial countries for specific deals. Britain and West Germany also negotiated projects that exchanged participation in technical and economic development for oil contracts.[25]

The major challenge, however, was from France. The French were pursuing both a national strategy, government-to-government deals for secure supplies of oil, and a European strategy, fostering a dialogue between the EEC and Arab countries. The Euro-Arab strategy did not make much headway. Although the EEC Council of Ministers approved a wide-ranging program of economic and technical cooperation in March 1974, the British vetoed a proposal for EEC negotiations with the Arab countries. Several meetings were held in an effort to institute a Euro-Arab dialogue, but they yielded no substantive agreements.[26]

As oil supplies became more available and prices stabilized in

[24]Lieber, *Oil and the Middle East War*, p. 30; Schneider, *Oil Price Revolution*, p. 255.
[25]Schneider, *Oil Price Revolution*, pp. 255–56.
[26]Lieber, *Oil and the Middle East War*.

mid-1974, the rush to secure direct contracts with producers slowed. Even the French, by the end of the year, were not eager for additional bilateral contracts, which had become in many instances more costly than oil purchased through the multinational oil companies. Bilateral agreements increasingly became a means to address balance of payments problems, with consuming nations looking for opportunities to market their goods among OPEC producers. In June 1974, for example, the French signed a ten-year, $4 billion agreement (later increased to $7 billion) with Iran for a package of economic and petroleum development projects. Oil was not directly involved in this deal.[27]

Although the threat of bilateralism was declining, U.S. executive officials continued to seek to put pressure on OPEC in order to moderate oil prices. Some success was achieved, not by collective restraint on the part of the consumer countries but through direct American pressure on Saudi Arabia. By September of 1974, however, Saudi efforts to continue a price moratorium had collapsed, and surpluses were countered not by a reduction of prices but by a lowering of production. At this juncture the United States again sought to bring consumer pressure on OPEC to moderate oil prices.

As before, the only leverage American officials were able to exert was the threat of collaborative actions by consumer nations to alter market forces through conservation and the development of alternative sources. In November 1974 Kissinger proposed new measures for cooperation among consumer nations. The first element of Kissinger's proposal was a set of safeguard measures to protect consumer nations from financial adjustment problems as well as to provide stand-by emergency oil-sharing arrangements. These measures were incorporated into the IEA. Second, Kissinger proposed a new set of conservation goals for the consuming nations. He proposed that the industrial nations take steps to reduce consumption of imported petroleum by 20 percent. Indicating that the United States was still intent on undercutting the pricing power of OPEC, Kissinger noted that his proposals could provide the conditions in two or three years that would make it "increasingly difficult for the cartel to operate."[28]

A year after his original London speech, then, Kissinger was still pursuing an international strategy that sought to unite the consumer nations. The original American position had envisaged immediate

[27]Schneider, *Oil Price Revolution*, p. 262.
[28]Quoted in *Business Week*, January 13, 1975, p. 67.

steps to confront OPEC and hinged on consumer nations not nego-
tiating bilateral agreements with oil producers. The new American
position was directed at a longer-term effort to alter the underlying
patterns of production and consumption. This shift in emphasis led
State Department officials to accept a broader dialogue between producers and consumers over the stabilization of prices.

BROADENING THE DIALOGUE

American officials now emphasized the need for an agreement between consuming nations and oil producers on a long-range floor
price or minimum selling price (MSP). The price of oil would come
down, these officials now argued, only with long-term efforts in conservation and the development of alternatives to oil. Efforts of this
sort would require massive investments by the industrial countries.
Such investments, however, could be jeopardized if OPEC manipu-
lated prices: oil prices could be brought just below the level necessary
to encourage the production of alternatives. To take the extraordinary investment measures needed to shift production and consumption patterns, therefore, the industrial nations needed to agree
among themselves, and with OPEC, to stabilize prices.[29]

The United States would facilitate a dialogue between consuming
and producing nations in order to establish stable international oil
prices. To protect Western investments in energy development, Kissinger proposed a common floor price for oil. "Paradoxically," Kissinger argued, "in order to protect the major investments in the industrialized countries that are needed to bring the international oil prices
down, we must ensure that the price of oil on the domestic market
does not fall below a certain price." He also proposed an international
consortium, to pool capital for energy investment as well as to conduct
joint research and development. With these agreements among consuming nations, Kissinger envisaged more constructive negotiations
with OPEC on prices, stable markets, and petrodollar recycling. The
bargain Kissinger sought to strike with OPEC turned on this logic:
OPEC should "accept a significant price reduction now in return for

[29]See discussion of this policy shift in Rustow and Mugno, *OPEC: Success and Prospects*, pp. 56–57. The scheme was championed by the State Department's assistant
secretary, Thomas O. Enders, and presented by Enders at a meeting of OECD representatives in April 1975. Its rationale is presented in Enders, "OPEC and the Industrial
Countries: The Next Ten Years," *Foreign Affairs* 53 (July 1975), 625–37.

93

stability over a longer period, or they can run the risk of a dramatic break in prices when the program of alternative energy sources begins to pay off."[30]

American officials began to campaign for a floor price on oil, but the proposal met with protracted debate in the IEA. Europe (except the United Kingdom) and Japan worried that an effective minimum selling price would thwart any long-term erosion in oil prices. They also feared that OPEC might construe such a proposal as political confrontation. With this consideration in mind, for instance, the Japanese foreign minister claimed that the price plan was "beyond the bounds of reason."[31]

Disagreements also broke out over what the floor price would accomplish. Behind the guarantee of a minimum selling price for oil, governments were to proceed to develop major new sources of energy. Yet the sources of energy to be developed and the effectiveness of the investment programs remained uncertain. Most important, conflict over the MSP hinged on the divergent gains that would accrue to energy-rich and energy-poor countries within the IEA. In effect, the MSP was a potential transfer of income to the energy-rich in return for the expectation of diffuse collective benefits: a better energy balance with OPEC. This conflict became more pronounced with the refusal of the United Kingdom and Norway to offer pledges of good faith concerning access by others to their domestic oil resources in return for agreement on the MSP.[32] Meanwhile, the credibility of the U.S. bargaining position was undercut by the continuing presence of American price controls on domestic oil.

The IEA eventually hammered out agreement on an MSP in September 1976. The agreement, however, was largely devoid of significance. The floor price for oil was set at $7 a barrel, far below prevailing world market levels. While proponents of the MSP could claim victory, few countries believed that prices would ever fall to such a low level, and though the agreement contained provisions for review of the program's implementation, such reviews have not taken place. Nor have there been efforts to move the price floor upward to reflect

[30]Address by Henry Kissinger, *News Release,* Bureau of Public Affairs, U.S. Department of State, February 3, 1975.

[31]Arabinda Ghosh, *OPEC, the Petroleum Industry, and United States Energy Policy* (London: Quorum, 1983), p. 155.

[32]See Peter F. Cowhey, *The Problems of Plenty: Energy Policy and International Politics* (Berkeley: University of California Press, 1985), pp. 253–54. See also Ann-Margaret Walton, "Atlantic Relations; Policy Coordination and Conflict: Atlantic Bargaining over Energy," *International Affairs* 52 (April 1976), 180–96; and Robert O. Keohane, "The International Energy Agency: State Power and Transgovernmental Politics," *International Organization* 32 (Autumn 1978), 939–40.

current costs of production.[33] An agreement may have been reached, but it could not mask basic disagreements among IEA members over the proper course of energy adjustment.

As the IEA continued to search for a meaningful agreement on a floor price plan, American officials shifted their strategy yet again. In May 1975 Kissinger placed oil price and supply issues on the agenda of the proposed dialogue between developed and developing nations.[34] Since the earliest moments of the October embargo, American executive officials had resisted efforts to place energy issues within a larger set of global economic negotiations. U.S. officials, while agreeing in principle to France's call for a dialogue with the oil-producing nations, argued that prior cooperation by consuming nations was necessary. In the wake of OPEC's success, however, government representatives from developing nations had begun to articulate an agenda for negotiations that would include other raw materials and broader international economic issues. Consequently, when the State Department's floor price initiative failed to attract support among consuming nations, the United States became interested in using the evolving North-South dialogue as a device for engaging OPEC on issues of price and supply.[35]

The American floor price proposal gave legitimacy to the demands of other commodity producers for cartel pricing agreements. In announcing the shift in American position in May 1975, Kissinger acknowledged that international agreements covering other raw materials would be considered on a "case-by-case basis." In this regard the American secretary of state proposed the creation of three separate commissions to cover energy, nonenergy raw materials, and economic development aid. This proposal met with resistance from OPEC governments; later in the year Kissinger, in a major UN address, proposed an elaborate set of global economic initiatives ranging from financial aid programs and commodity agreements to a new, special $10 billion IMF lending facility to stabilize the export earnings of

[33]Cowhey, *Problems of Plenty*, p. 254.

[34]The willingness of the United States to participate in a preparatory conference between producers and consumers was linked to an American-European understanding that steps would be taken to increase consumer solidarity. The change in the U.S. position, however, was not dependent on the promise of specific agreements within the IEA. See Walton, "Atlantic Bargaining over Energy," p. 193. The new American position was outlined by the secretary of state in two speeches on May 27 and 28, 1975. See Clyde H. Farnsworth, "Kissinger Offers New U.S. Aid Plan," *New York Times*, May 29, 1975, pp. 1, 19. Domestic debate over the merits of engaging in the so-called North-South dialogue is discussed in Geoffrey Barraclough, "Wealth and Power: The Politics of Food and Oil," *New York Review of Books*, August 7, 1975.

[35]Schneider, *Oil Price Revolution*, pp. 268–69.

developing countries. These various programs were to be financed jointly by OPEC and the industrial countries. Talks got under way in Paris as representatives of developed and developing countries met at the Conference on International Economic Cooperation (CIEC).[36]

These efforts to manage commodity markets internationally did not, however, provide any headway on the American effort to induce OPEC to moderate or reduce oil prices. The United States opposed commodity schemes involving the maintenance of prices at levels necessary to achieve significant transfers of wealth. Nor would the consumer nations accept OPEC's proposal for the indexation of oil prices to the price of producer goods. The demise of the U.S. international adjustment strategy came in November 1976. On the eve of a meeting of the CIEC, Kissinger sent word to the representatives of other industrialized countries conceding that "there is no negotiable package which the industrialized countries could accept and which would also present sufficient inducement to OPEC to refrain from a substantial oil price increase over several years, given the lack of leverage by consumers over oil prices."[37] American efforts to orchestrate a united response to OPEC by consumer countries, which had evolved over the preceding three years but had constantly sought the moderation or rollback of cartel prices, were at an end.

THE IEA AND THE SECOND OIL CRISIS

In the three years following the October 1973 embargo American administration officials pursued an international offensive strategy that sought to unite the consumer nations and, by so doing, develop the economic and political leverage to return oil prices to more moderate and stable levels. At the center of consumer unity was to be the International Energy Agency. The IEA, however, did not play the role that American officials had envisaged. Governments in Europe and Japan resisted efforts to use it as a vehicle to confront OPEC. The IEA did manage to develop a more modest mission that revolved around agreements on emergency oil sharing and energy information sharing. In effect, the agency came to represent consumer country efforts to develop international *defensive* measures to guard

[36]The background and evolution of international bargaining over commodity price stabilization schemes and other issues within the North-South dialogue are covered in Robert L. Rothstein, *Global Bargaining: UNCTAD and the Quest for a New International Economic Order* (Princeton: Princeton University Press, 1979).

[37]*New York Times*, December 14, 1976. See also Rothstein, *Global Bargaining*, p. 47 n. 16.

against future emergencies. Yet as we shall see, in 1979 cooperation between the consuming nations remained, for the most part, elusive.

A central mission of the IEA was the development of an emergency oil-sharing agreement. The task was to develop an agreed set of procedures between the major international oil companies, the IEA, and member-governments for the sharing and management of oil supplies at moments of extreme shortage. The IEA also began to develop the organizational capacity to monitor international oil markets and to set aggregate oil-import targets for member-countries as a whole.[38]

The stand-by agreements and monitoring activities of the IEA did not significantly moderate the scramble for supplies in the midst of the 1979 oil crisis. This crisis, triggered by the cutback of Iranian production, resulted in a reduction of about 2 million barrels per day on international oil markets in late 1978. Although the aggregate loss of production was modest, approximately 6.3 percent of the noncommunist world total, the rush by governments and companies to secure supplies was even more intense than in 1973–74. International oil companies, independent refiners, and governments all acted to increase their stocks. In the face of uncertainty, demand for oil on the spot market rose quickly, pushing prices to extraordinary levels, far out of proportion to actual shortages.[39]

As in the earlier oil crisis, prices were driven by short-term, competitive responses to very small shortfalls in production. The Iranian cutback, which caused prices to climb on the spot market, was followed in January 1979 by the announcement that Saudi Arabia would reduce production to meet a ceiling set by the government. Production dropped only slightly (by approximately 500,000 barrels a day for several weeks in February) and was restored in March, but the effect on prices was dramatic. The spot price for Saudi Arabian crude rose 64.5 percent, from $19.09 to $31.40 a barrel, during February and dropped back to $25.04 a barrel in late March[40] (see Table 4).

[38]Keohane, "International Energy Agency," pp. 929–52, and Keohane, *After Hegemony*, pp. 224–37. Other discussions of IEA activities include Cowhey, *Problem of Plenty*, chap. 8; Walton, "Atlantic Relations"; Wilfred Kohl, "The International Energy Agency: The Political Context," in J. C. Hurewitz, ed., *Oil, the Arab-Israel Dispute and the Industrial World* (Boulder, Colo.: Westview, 1976); Mason Willrich and Melvin A. Conant, "The International Energy Agency: An Interpretation and Assessment," *American Journal of International Law* 71 (April 1977), 199–223.

[39]See Daniel B. Badger, Jr., "The Anatomy of a 'Minor Disruption': Missed Opportunities," in Alvin L. Alm and Robert J. Weiner, *Oil Shock: Policy Response and Implementation* (Cambridge, Mass.: Ballinger, 1984), pp. 33–53. See also Badger and Robert Belgrave, *Oil Supply and Price: What Went Right in 1980?* Energy Paper no. 2 (London: Royal Institute of International Affairs, 1982), p. 107; Keohane, *After Hegemony*, pp. 226–28.

[40]Verleger, *Oil Markets in Turmoil*, pp. 34–35.

Table 4. Spot market and official oil prices, January 1978–April 1980 (cost per barrel in U.S. dollars)

	Spot Market Value	Official Contract Price
1978 Jan.	$13.71	$14.20
April	14.13	13.95
July	14.14	13.90
Oct.	15.42	13.95
1979 Jan.	19.71	14.73
April	27.65	18.42
July	34.35	23.41
Oct.	35.79	24.19
1980 Jan.	38.32	33.10
April	32.27	35.65

SOURCE. *Petroleum Intelligence Weekly,* Special Supplement, February 2, 1981.

Although supply-side changes were small, they triggered demand-side actions that were large and consequential. The uncertainty created by the suddenness of the cutbacks and the relatively low inventories that oil users possessed at the moment of the disruptions encouraged competitive bidding for supplies. At the moment that supplies were tightening up, short-term demand for those supplies increased.

Efforts by individual buyers to increase inventories escalated prices out of line with the loss of production. Indeed, the shortfall in production was very brief. By the third quarter of 1979 world oil production was actually higher than it had been the previous year, and consumption was only marginally higher. Iranian production did decline in late 1978 and in 1979, but other Middle Eastern oil producers, particularly Saudi Arabia (in the third and fourth quarter of 1979) and Iraq, raised production over the previous year's level.[41]

Efforts by the IEA and its member-governments to address the collective dilemma of competitive bidding were unavailing.[42] The IEA urged in March 1979 that consumer countries cut their imports by 2 million barrels a day. Yet individual governments were not held to specific targets, and the incentives for demand restraint were not

[41]U.S. Department of Energy, Energy Information Administration, "1979 International Energy Annual" (Washington, D.C., August 1980), p. 11.

[42]Indeed, many actions by governments tended to increase market pressures. The United States, in May 1979, imposed a $5 per barrel subsidy for distillate imports. This subsidy, imposed without warning, was roundly criticized by IEA member-countries and immediately drove up prices on the spot market. At the same time, oil stocks under the control of the governments of the consuming nations could have been released on to commercial markets, mitigating the competitive scramble for supplies. See Badger, "Anatomy of a Minor Crisis."

strengthened. Imports by IEA countries in fact increased in 1979, by about 1 percent over 1978 levels. Nor was the IEA's emergency sharing agreement of use. That agreement, to redistribute oil supplies to IEA members with more than a 7 percent shortfall of supply, was designed to protect countries from selective embargoes. In 1979 shortfalls for individual countries hinged on such circumstances as the existence of price controls and stockpile arrangements. The sharing of oil would at least partially have rewarded countries with less efficient import policies.[43]

The consuming nations also used the occasion of their annual economic summit meeting, held in Tokyo in June 1979, to agree to country targets for oil imports. Prior to the summit the French government, in contrast to its actions in 1974, called for quantitative limits on oil imports, coordination of emergency stockpiling programs, and joint control of prices on the spot market. At a meeting of the European Council, however, government representatives could agree only on the monitoring of the spot market and refused to commit themselves to national import targets. During the Tokyo summit and in later negotiations, Japan and West Germany forcefully opposed controls on spot market purchases and national import levels. Japan was eager to avoid international obligations that might prevent it from finding new supply contracts, which it needed as contracts with the major oil companies terminated. West Germany was suspicious of controls on the market and was convinced that the inevitable cheating on any agreement would be sufficient to drive prices up anyway.[44]

Agreement at the Tokyo summit on national targets to limit oil imports was possible only when pledges gave governments large margin for discretion. A compromise was eventually concluded: the United States, Canada, and Japan agreed to national ceilings, and the European governments agreed to reach individual national targets later. Following the summit the United States continued its attempt to persuade other consuming nations to share the burden of supply reductions.[45] Yet the United States itself was having difficulty reducing imports, and the target agreements were of limited effectiveness in coping with the immediate problems of OPEC supply reductions.

[43]See Keohane, *After Hegemony*, pp. 228–29.

[44]Cowhey, *Problems of Plenty*, p. 274.

[45]See George de Menil and Anthony M. Solomon, *Economic Summitry* (New York: Council on Foreign Relations, 1983), pp. 28–29. Subsequent U.S. efforts took place primarily through the IEA. See "A Push for Unity against OPEC," *Business Week*, December 3, 1979, pp. 36–37.

The IEA and member-governments were ill-equipped to deal with the problem of collective action. As a result governments, oil companies, and marketers followed the incentives of the moment. Short-term demand had the effect of doubling oil prices. The OPEC cartel followed the spot market, moving official prices higher as well. After the disruption ended and supplies had been stabilized, the price of oil remained at the higher level.[46]

CONCLUSION

The rise in oil prices in 1973–74 and in 1979 began with the curtailment of a portion of OPEC production. The manner in which consumer nations responded to these disruptions in supply had a decisive impact on the magnitude of the price increases. The scramble by state officials and oil companies to ensure supplies pushed prices higher and reinforced the market position of the oil cartel. Opportunities for cooperation existed at both moments of crisis, but cooperation remained elusive.

The interests of states in a multilateral strategy to cope with oil shortages diverged, and these divergences were related to differences in the capacities of those governments to pursue other strategies. The interest in and fate of international strategies of adjustment can be understood when those strategies are juxtaposed with the other opportunities states had for action. In the American case, other opportunities were not immediately available. Price controls constrained the ability of American executive officials to impose higher costs of imported oil on the domestic economy. Moreover, foreign policy officials could conduct an international strategy of adjustment without getting immediately involved in the larger and unwieldy decision-making process. At the same time the international position of the United States provided opportunities to seek international adjustment solutions—solutions that accorded with a broader set of hegemonic political objectives. Other consuming nations, however, were able to pass

[46]A third oil supply disruption, precipitated in 1980 by the Iran-Iraq war, did not have the same effects on prices as the two previous disruptions even though the fall in the world production involved was actually greater than the production losses of 1979. Over a two-month period, world oil production fell by 3.6 million barrels a day from 59.5 to 55.9 million barrels a day. Yet oil prices did not increase as in the earlier crises. Although an increase in Saudi production was important in offsetting these losses, Verleger argues that the most important factor was the high level of consumer stocks. "Stocks were high in the fall of 1980, and oil companies probably welcomed the opportunity to work them off." *Oil Markets in Turmoil*, p. 38.

prices through their economies and were better equipped to pursue separate deals with oil producers. The construction of multilateral schemes to share the costs of adjustment and to moderate price increases, at least in the 1973–74 crisis, was less pressing than other objectives: access to supply and favorable diplomatic relations with Arab states. Moreover, while U.S. officials had stronger incentives and a favorable international position from which to pursue an international offensive strategy, their inability to adjust to higher prices domestically prevented them from satisfying their international obligations. The persistence of price controls indirectly encouraged the growth of oil imports. Even if other nations had agreed to a collaborative program of import restraint, the United States could not have lived up to its part of the bargain.

This argument is not a complete theory of the determinants of international collaboration. The industrial importing nations confronted divergent international and domestic constraints and opportunities, which in turn influenced national preferences for specific forms of international agreements. But states with different international and domestic capacities may still participate in multilateral cooperative schemes. The French, for example, were much more interested in coordinating national oil-import levels during the second oil crisis than during the first. Moreover, some of the conflicts between the industrial nations over energy adjustment were rooted in differences only indirectly related to state capacity. Lines of conflict over the MSP, for instance, were primarily drawn between countries with and those without significant indigenous sources of energy. I focus here, that is to say, on a particular set of constraints and opportunities—variables that cannot explain the entire course of cooperation and noncooperation in energy adjustment during the 1970s. These variables are helpful, however, in explaining important elements of the international struggle over energy adjustment: the initial interest of American officials in an international solution to OPEC petroleum pricing; the opportunities for other industrial nations, particularly France, to "exit" from these multilateral schemes; and the inability of the United States to fulfill its leadership role by controlling domestic imports of petroleum.

It is difficult to disentangle the conflicts over energy among industrial nations in the 1970s from larger foreign policy and economic circumstances. American international strategy sought to advance goals beyond the immediate problem of energy adjustment. French opposition to that strategy was also wrapped up in a larger and well-established objection to American leadership. Differences among states concerning the nature of relationships with the countries of the Middle East served to widen the breach between Washington and other Western capitals.

The international strategies that American officials articulated and reformulated between 1973 and 1976 were only a subset of a larger variety of strategies available to states. U.S. officials were drawn to the multilateral approach because, if successful, it would have the virtue of lowering and distributing the costs of adjustment across the industrial world—an option all the more attractive for executive officials ill-equipped to impose costs on their own economy. At the same time Secretary Kissinger was intent on removing OPEC from a position of prominence in the international system. OPEC had challenged the West with higher oil prices. With higher energy costs came slower growth, export competition, greater bilateralism, and unprecedented demands by developing countries for fundamental reform of international economic relations—all events that ran afoul of the principles upon which American postwar leadership was based.

The failure of the international collaborative proposals advanced by the State Department during the 1973–74 crisis was rooted in the "exit" by the Japanese and the Europeans, particularly the French, from those agreements. French opposition to the American strategy was informed by a variety of factors. The larger issues of political leadership within the Western alliance, divergent foreign policy positions on the Middle East conflict, and European regional goals all formed part of the backdrop for French intransigence in the face of American multilateral initiatives.[47] Beneath these political considerations were divergent national economic and political capabilities. While imports in the United States continued to rise, the French government in 1975 moved to limit annual oil imports to about 10 percent below the level of the previous year.[48] The Ford administration, in the same year, sought to limit oil imports by use of a tax; this import fee was an attempt to pressure Congress into decontrolling oil prices, but the effort did not succeed and imports continued to rise. The French were also well-equipped to negotiate bilaterally with oil producers. Through state-owned oil companies the French government could negotiate secure contracts of oil directly, even in the face of overall shortages and dislocations in world markets. Because of

[47]The most telling presentation of the French view on American international energy strategy came in the remarks by Minister of Foreign Affairs Michel Jobert at the 1974 Washington Energy Conference. See Embassy of France, "Statement by His Excellency Michel Jobert" (Washington, D.C., February 11, 1974).
[48]Rustow and Mugno, *OPEC: Success and Prospects*, p. 55.

these and similar differences in political objectives and national capabilities, the Europeans and the Japanese did not find compelling the multilateral schemes pursued by the United States during the 1973–74 oil crisis.

Finally, the United States itself was not able to develop the domestic responses that would have given its international strategies more credibility.[49] In particular, the United States was not able to curb its own consumption of imported oil.[50] European demand for oil dropped from an average of 15.4 million barrels a day in 1973 to roughly 13 million barrels a day in 1978, the year preceding the second oil shock. During the same period Japanese consumption of imported oil also fell, if slightly less. U.S. oil consumption, however, rose substantially: oil imports went from an average 6.3 million barrels a day in 1973 to 8.2 million in 1978 (and rose again in 1979).[51]

American executive officials attempted to redress the nation's problems of adjustment by confronting OPEC directly over the pricing issue. Other industrial importing countries, with different capabilities and policy objectives, found alternative responses, and in the end the United States was also forced to look for other responses. When the oil shortages and dislocations of 1973–74 receded, the price consequences remained, and the United States found itself looking for domestic (as well as international) means to adjust to the new reality.

---

[49]Kissinger noted this failure. "When I was in office, we helped develop the International Energy Agency, which was supposed to bring together the industrial democracies in a program of conservation, alternative supplies and emergencies against embargoes. [He went on to argue that] I think it is safe to say that this agency has not been able to achieve the goals set for it, largely because the United States did not carry out its share of the necessary conservation." U.S. Senate, Committee on Energy and Natural Resources, Subcommittee on Energy and Regulation, "Status of Federal Energy Conservation," Hearings, April 4, 1977, 95th Cong., 1st sess., p. 5.

[50]An analyst writing in 1976 noted the influence of domestic constraints on the waning vigor of American energy diplomacy. The retreat of U.S. international efforts "lies in the relatively weak position of the United States regarding the development of its domestic energy policy. Domestic linkages, the political constraint which stops governments from purusing a policy of restraint through pricing, have been particularly apparent in the United States." Walton, "Atlantic Bargaining over Energy," p. 195.

[51]*BP Statistical Review of World Energy*, June 1984, p. 16.

CHAPTER FIVE

The Limits of State Building

> Only the expert knowledge of private economic interest groups in the field of "business" is superior to the expert knowledge of the bureaucracy. This is so because the exact knowledge of facts in their field is vital to the economic existence of businessmen. Errors in official statistics do not have direct economic consequences for the guilty official, but errors in the calculation of a capitalist enterprise are paid for by losses, perhaps by its existence. The "secret," as a means of power, is, after all, more safely hidden in the books of an enterpriser than it is in the files of public authorities. For this reason alone authorities are held within narrow barriers when they seek to influence economic life in the capitalist epoch. Very frequently the measures of the state in the field of capitalism take unforeseen and unintended courses, or they are made illusory by the superior expert knowledge of interest groups.
>
> Max Weber, 1946

The initial American response to the 1973 price shocks was an attempt led by the State Department to forge a common front among Western industrial nations. As these international efforts increasingly were frustrated, executive officials sought a second avenue of adjustment that entailed redefining the responsibility of the state for domestic energy planning and production. In effect, various sets of executive officials and politicians attempted to break out of the institutional constraints that left the state with few instruments or capacities within the energy sector.

In this chapter I analyze the various attempts by government officials to redefine the state's role in the energy sector. I look at the problem of information gathering and competence in the energy area; at the attempts to impose a government presence in the energy sector through a federal petroleum corporation; at attempts to create mechanisms for energy financing; and at the establishment of the synfuels corporation. These efforts were largely unsuccessful. At each juncture, and in ways reminiscent of failed state-building proposals of earlier historical periods, the proposals were blocked—by Congress, by interest groups, and by divisions within the executive branch itself.

The failure to build additional state capacity to cope with the energy dilemmas of the 1970s was not due to the absence of imaginative proposals. Initiatives to expand the state's role quickly became part of the politics of energy adjustment. These initiatives sought both to create new realms of government planning capacity and to establish a direct state presence in the energy sector.

Three important proposals came from various sets of executive officials. The earliest effort, during the Nixon administration, sought to increase the energy information and expertise available to policy makers. This was Project Independence. Another was the Ford administration's $100 billion Energy Resource Finance Corporation, which attempted to create a capacity for the direct financing of energy production and exploration. The final proposal, the Carter administration's Synfuels Corporation, drew on similar initiatives advanced in earlier decades and had a legacy of congressional sponsorship. This earlier history helps explain why the Synfuels Corporation was the only proposal to be implemented. A final proposal, to create a state-owned petroleum corporation, came from several members of Congress.

The constraints on state building can be understood at several levels. These proposals provoked opposition from a wide variety of interest groups and officials. Executive officials and congressional politicians, maintaining a commitment to market mechanisms, resisted them, and groups within the energy industry itself actively registered their opposition as well. Opposition within the executive policy establishment undercut passage on several occasions. In the policy battles that were waged, the proponents of an expanded state role in energy planning, financing, and production consistently lost.

It is important to situate these policy stuggles within a broader historical and institutional setting. In the first place, the scarcity and fragmentation of bureaucratic expertise and operational capabilities provided few bases on which to build new government powers and responsibilities. Proponents of the various state-building designs were scattered across the federal establishment and only infrequently enjoyed strong presidential support. Moreover, the earliest state-building efforts centered on the development of analytic or informational capacities, without which ambitious programs that involved a direct government role in energy financing or production could not proceed. Indeed, the struggle over a greater state role within the energy sector hinged in many respects on disagreements over the nature of energy markets, particularly supply and demand elasticities.

PROJECT INDEPENDENCE

On October 17, 1973, the Arab oil producers announced their boycott of American and Dutch markets. On November 7 President Richard Nixon addressed the nation on the subject of the emerging energy crisis. He announced immediate steps to help mitigate fuel shortages and stimulate domestic energy production. He also announced a large, long-range program. Project independence, drawing on the spirit of the Manhattan and Apollo projects, had the goal of developing "the potential to meet our own energy needs without depending on any foreign energy sources" by the end of the decade.[1] Project Independence articulated a goal more than it set out concrete proposals. The program was to include both technological and non-technological initiatives, incorporating the proposals Nixon had made in June 1973 for a massive program of energy research and development. Short on specifics, the project held out the intention of a more inclusive energy policy, giving research and development a role in this larger strategy. The key proposal was a major, wide-ranging analytic study to guide national spending and allocative decisions.

The Project Independence Evaluation System

While statutory authority for the creation of a Federal Energy Administration was pending, and while the larger proposal for a Department of Energy and Natural Resources was blocked in Congress, Nixon established a small Federal Energy Office to begin initial policy analysis. Although organizational changes were constant, officials from executive agencies were brought together in the FEO, and later the Federal Energy Administration, to launch analytical studies. The major study, known as the Project Independence Evaluation System (PIES), was an attempt to create a broad-based and systematic model of energy demand and supply, with the intention of making projec-

---

[1]President Richard M. Nixon, "Address on the Energy Emergency," November 7, 1973, in *Presidential Energy Statements*, Committee Print, Committee on Interior and Insular Affairs, U.S. Senate (Washington, D.C., 1973), p. 86.

tions of the nation's energy future. Led by John Sawhill, recruited from the Office of Management and Budget, and Eric Zausner, recruited from the Department of Interior, the system became the first full-scale energy study undertaken within the American government. The task was to evaluate the nation's energy problems and specify alternative policy options.[2]

The Project Independence study was set in motion under severe time constraints. Its analysis was to undergird a national energy program, and the pressure for such a program was very great. The report was to be completed by the fall of 1974. With a premium on time, the project gathered experts and specialists from all quarters of the federal establishment. Represented in the project, for example, were officials recruited from the Office of Management and Budget, the Environmental Protection Agency and the departments of Labor and Commerce. Altogether officials from over twenty-five agencies were involved, and their recruitment was seen as a move to increase both the coordination of the subsequent energy policy and the acceptance of analytic modeling within the bureaucracy.[3]

As model building proceeded, analysts began to play down the political goal of complete self-sufficiency. At the same time they attempted to broaden the project so as to provide projections on the implications of the wide array of possible presidential energy initiatives. Although their original mandate was to produce specific policy recommendations, the analysts soon redefined their mission more generally; as Sawhill remarked, "it was not the intent of this specific report to recommend a total national energy program. Energy policy is a complex issue and the formulation of appropriate Federal programs and policies can entail many paths and choices. The Project Independence report was intended to present a comprehensive framework within which the public, Congress, and the Executive Branch could evaluate individual issues."[4] Zausner notes that the shift in mission from policy recommendation to the creation of a general analytic framework resulted from a desire to protect the authoritative nature of the enterprise. His central objective was an "institutional upgrade" in energy information and analysis. If the project moved

[2]See Federal Energy Administration, *Project Independence: A Summary* (Washington, D.C.: GPO, November 1974); Martin Greenberger, *Caught Unawares: The Energy Decade in Retrospect* (Cambridge, Mass.: Ballinger, 1983), p. 110.

[3]See Thomas H. Tietenberg, *Energy Planning and Policy: The Political Economy of Project Independence* (Lexington, Mass.: Lexington Books, 1975), p. 46.

[4]John Sawhill, "Project Independence," Hearings before the Committee on Interior and Insular Affairs, U.S. Senate, 93d Cong., 2d sess., November 21, 1974, p. 164.

immediately to recommend policy, subsequent attacks could undermine its larger purpose.[5]

The PIES project snowballed into a massive administrative under-
taking involving over five hundred analysts and numerous task
forces. Project Independence staff conducted hearings across the
country, attempting to find areas of consensus. Also attached to the
project was an Advisory Committee, which drew its twenty-eight
members from industry, labor, and state and local government, as
well as public interest groups.[6] The actual significance of the hearings
and the committee is disputed. Nonetheless, these additional activities
do reflect the centrality FEA officials understood their Project to
possess for the eventual decision on a national energy policy.

The project involved many individuals, but the bulk of the report
was generated by a core staff of several dozen. The task forces,
Zausner notes, were designed to "bring other administrative officials
into the tent." The task forces were not crucial to the analytic work
but rather were used for two purposes. One was to provide data
resources. Even as late as 1976, after all, twenty-three executive departments and independent agencies operated 238 major energy
data–gathering programs. Zausner notes, for example, that the Interior Department had important coal data but did not have the analytical sophistication to use them. The task forces mobilized this information. Second, the project staff sought to bring other administrative
officials into the enterprise at an early stage so as to protect the subsequent analytical product. The typical method of bureaucratic attack is
to challenge the assumptions of a study, and so project staff attempted to get other bureaucratic players to agree on assumptions
early on, coopting them and thereby foreclosing potential conflict.[7]

The final report was produced in November 1974, one year after
Nixon had declared Project Independence. It provided models of
energy demand and supply, as well as larger models that related
energy shifts to macroeconomic, environmental, and other variables.[8]

[5]Interview, Eric Zausner, June 26, 1984.
[6]See Neil De Marchi, "Energy Policy under Nixon: Mainly Putting Out Fires," in
Craufurd D. Goodwin, ed., *Energy Policy in Perspective: Today's Problems, Yesterday's Solutions* (Washington: Brookings, 1981), pp. 458–66; Greenberger, *Caught Unawares*, p.
111; and Joel Havemann and James G. Phillips, "Energy Report/Independence Blueprint Weighs Various Options," *National Journal Reports,* November 2, 1974, p. 1636.
[7]Interview, Eric Zausner, June 26, 1984. The count of data-gathering programs is
from General Accounting Office, *Report to the President and the Congress: Performance
Evaluation of the Energy Information Administration* (Washington, D.C., June 15, 1984), p.
1–1.
[8]Federal Energy Administration, *Project Independence Report* (Washington: GPO,
November 1974).

It generated both a massive substantive study of energy demand and supply and a complex, computer-based set of models to guide subsequent analysis of policy options. This analytical framework marked a new level of sophistication in energy policy analysis. As one FEA official argued, "the FEA model is an order of magnitude more sophisticated than anything done before. Whereas we had sweeping generalities before, we can now look at regional energy flows in detail. . . No matter what happens to the policy scenarios laid out in the blueprint, we've taken a major step forward in developing analytical tools for looking at energy problems."[9]

The project's list of various steps to promote domestic energy production tended to underline constraints on production. The report argued, for example, that the acceleration of nuclear power plant construction would not significantly reduce oil imports. Nor would a massive effort to produce synfuels, bypassing important research steps, be very practical or efficient. Although the report cast doubt on the possibility of significant short-term gains in energy independence, it stressed that increases in self-sufficiency would have an impact on world oil prices. The study concluded that with a reduction of U.S. imports, OPEC would find it hard to maintain a price of even $7 a barrel. But price reductions, it noted, would harm new domestic energy investment, perhaps requiring government price guarantees and supports.[10]

Project Independence and the Policy Process

The PIES analytical models had more than a modest impact in government during the Ford administration. As one study notes, "for the first time, the government appeared to have its own source of energy information and analysis—reasonably integrated, increasingly rationalized, and operated and maintained in-house."[11] The chief architect of the modeling project, FEA assistant administrator Zausner, also noted the importance of the massive study. "Until the blueprint," Zausner claimed, "we really had a clean slate."[12] The implication was that the project reflected a dramatic departure in government capacity for planning and analysis. Zausner has since noted that there was a "tremendous lack of information and analysis"

[9]Unidentified official quoted in Havemann and Phillips, "Energy Report," November 2, 1974, p. 1640.
[10]*Project Independence Report*, November 1974, pp. 8, 15.
[11]Greenberger, *Caught Unawares*, p. 112.
[12]Quoted in Havemann and Phillips, "Energy Report," November 2, 1974, p. 1636.

prior to Project Independence. Nixon's project opened up the opportunity to "build a body of experts," to build an analytical model, and "put some important facts on the table."[13]

The models were used at various points to calculate the impact of oil decontrol on inflation and energy supply. In December 1974, while President Ford was preparing new energy proposals, the FEA modeling system was brought into the process. The FEA prepared briefing materials for a Camp David policy meeting during this period. "The Project Independence analysis," one author argues, "was the major source of information consulted in preparing these briefing books. In retrospect the role it played was to point out that this historical policy package was not sufficient to meet the administration's goal either in the immediate future or in the long run, if world prices were to fall."[14] Another account suggests that the Project Independence analysis played a less central role in those December policy deliberations. Some of the issues brought before Ford in December "specifically call[ed] for critiques of the Project Independence Report or for development of additional data."[15] Zausner indicates that the project analysis "dominated the process"; the project did not have all the information Ford needed for decisions, but the Camp David policy review "spoke to" the project findings.[16] Indeed, as the Ford energy package took shape, initiatives came directly or indirectly from the PIES analysis, among them financial support for the electric utilities industry, the establishment of a strategic petroleum reserve, and tax measures and voluntary standards to promote conservation.

Yet the PIES analysis did not have a sustained impact on administrative decision making, thus falling short of the hopes of its authors. The influence of PIES analysis and its FEA sponsors did not extend far within the administration. The project had brought into its ranks specialists from all reaches of the executive bureaucracy, but the report's analysis and methodology did not follow those officials back to their respective agencies. Because of the short period of time available to the project, the final report did not circulate widely within the administration before publication. Many agencies had only a week to review the document. Thus many administrative units may have had a substantive stake in the report's findings, but they did not have

[13]Interview, Eric Zausner, June 26, 1984.
[14]Tietenberg, *Energy Planning and Policy*, p. 88.
[15]Joel Havemann, "Energy Report/Federal Planners Study Ways to Cut Reliance on Imports," *National Journal Reports*, December 14, 1974, p. 1863.
[16]Interview, Eric Zausner, June 26, 1984.

an organizational stake in its production. As a result, the modeling project was not well received in the executive establishment.[17]

The problem of bureaucratic agreement was tied up with a variety of battles over turf. The Project Independence report was in effect a claim by the Federal Energy Administration to a dominant role in policy making. For the FEA to claim the preeminent role, it had to have the best analytical unit in the executive establishment. "He who has the data base and the analytic model," Zausner argued, "has the bureaucratic muscle."[18]

The influence of the model and analysis depended heavily on the status of Project Independence officials in the policy process, a status that changed frequently under the Ford administration. FEA officials, one author notes, "were not effective in integrating it [PIES] into the policy-making process." During the formulation of President Ford's first major speech on energy and the economy, in early October 1974, for example, the Federal Energy Administration submitted options based on Project Independence analysis, among them a proposal for a gas tax. However, the FEA's chief, John Sawhill, was not well placed in the administration (and later that month was forced to resign). Consequently, FEA and the PIES analysis did not play a major role in Ford's October energy proposals.[19] Indeed, Project Independence officials did not present their November report directly to the president, as they had anticipated doing. Instead, top officials instructed FEA to submit the report to the Energy Resources Council. This executive council, under the direction of Rogers Morton, had maneuvered at FEA's expense to become the chief body for policy recommendation. Morton's unsympathetic attitude toward the Project Independence analysis removed that material and its officials from the direct formulation of policy. Later, with the appointment of Frank Zarb as FEA chief, the analysis was brought into high-level policy planning more frequently, but the influence came from Zarb, an administrative insider, not from the independent authority of the PIES analysis. During this period it was the FEA that "staffed" the Energy Resources Council, where Zarb was executive director. During Zarb's tenure, as a result, all the important decision memoranda went through Zausner and the FEA.[20]

[17]Havemann and Phillips, "Energy Report," November 2, 1974, p. 1653; Tietenberg, *Energy Planning and Policy*, pp. 46–47.
[18]Interview, Eric Zausner, June 26, 1984.
[19]Tietenberg, *Energy Planning and Policy*, pp. 63, 87.
[20]Edward Cowan, "Zarb Is Praised as Man in Middle on Oil Price Dispute," *New York Times*, July 23, 1975; interview, Eric Zausner, June 26, 1984.

After Zarb's departure, however, Rogers Morton again removed project officials from policy deliberations.[21] Finally, with the departure of key FEA officials in 1976, the PIES operation began to lose what role it had achieved in the policy process. The analytical framework survived into the Carter administration, and PIES evaluation and forecasting models were used in the preparation of the first Carter energy program. By this time, however, the reputation of PIES as an authoritative device had diminished considerably.

The Elusiveness of Authoritative Information

Attempts to establish an authoritative body for energy analysis and information continued beyond the failure of Project Independence. In August 1976 Congress passed the Energy Conservation and Production Act, authorizing the establishment of a separate information and analysis unit within the FEA. Congress intended to separate analysis from the staff and policy-making units within the agency: the act called for the information bureau to be "insulated from FEA's role in formulating and advocating national policy."[22] When the Department of Energy was established in 1978, this independent bureau became the Energy Information Administration.

The legacy of Project Independence, therefore, was an initially ambitious effort to bring new analytical capacities to the center of policy formation. The analysis concerned had its limitations, however. The data remained quite aggregated, and many policy participants found the direct relevance of the analysis to policy elusive. Project Independence was the first major effort to develop in-house expertise on energy supply and demand and other macroeconomic variables, yet the modeling effort and its experts were unable to remain at the center of policy. Challenges to the authority of PIES followed its initial report—challenges from Congress and from various quarters within the administration. When an agency—the Energy Information Administration—finally was established, its organizational mandate consigned it to the peripheries of the formulation of national energy policy.

The FEA controlled most phases of information and analysis prior to the establishment of the Department of Energy. Its three important divisions were data collection, analysis, and policy. The creation

[21]Edward Cowan, "Ford Calls Aides to Weigh Energy and Economy," *New York Times*, December 26, 1974, p. 45.

[22]*Energy Conservation and Production Act,* U.S. Public Law 94-385, enacted August 14, 1976, secs. 51a, 142; discussed in Greenberger, *Caught Unawares*, p. 116.

of the Energy Information Administration separated data collection and analysis from policy formulation. The EIA's information responsibilities, one study notes, "included carrying out a unified program to collect, process, and publish data and information relevant to energy resource reserves, production. demand, and technology."[23] Using as its model the Bureau of Labor Statistics, Congress successfully established the information function independent of administrative policy making. In effect, executive and congressional officials alike had suspicions about energy information and analysis, and as a result created a new institution in such a way as to make it peripheral to the policy process.

The goal of an apparatus that would generate authoritative planning and analysis, and be located in a powerful and centralized administrative unit, remained elusive. Even basic information about energy supplies and flows was in dispute.[24] This absence of authoritative knowledge was reflected in a sense of uncertainty on the part of officials within the policy process. One senator expressed this sentiment in the fall of 1974: "When we first started debating this problem, after the Middle East crisis, one of the most serious problems we encountered was a tremendous lack of information regarding the facts. There were figures quoted as to how much we were importing and, at that point in time, figures ranged from as little as 15 to as much as 40 percent. As far as true information to our future potential supplies, I understand that is difficult to come by, but it was almost impossible to even gather the basic facts to base a national policy on."[25]

Administrative and congressional debate over energy policy after 1973 was manifestly political and centered on substantive policy options. But it was fed by great uncertainty both as to the availability of data and analysis and as to their source. At the very outset government was highly dependent on the energy industries themselves for information. One report in early 1974 noted: "By and large, the

[23]General Accounting Office, *Report to the President and the Congress: Performance Evaluation of the Energy Information Administration* (Washington, D.C., June 15, 1984), p. 1–1.

[24]Basic information was not available on oil flows in the midst of the October 1973 embargo. For example, after the embargo William Simon asked Eric Zausner, in charge of data and analysis in the Interior Department, for figures on current oil import levels. The one agency with energy information was the Bureau of Mines, in the Interior Department. That bureau reportedly told Simon it would have available data on import levels for 1971 in another six months. Interview, Eric Zausner, June 26, 1984.

[25]Statement by Senator Pete V. Domenici, "Project Independence," Hearings before the Committee on Interior and Insular Affairs, U.S. Senate, 93d Cong., 2d sess., November 21, 1974, p. 5.

industry's own figures—unverified by independent audit—are the only ones presently available to guide federal energy policy decisions."[26]

The government's ability to gather energy information had evolved piecemeal prior to the 1970s. At the moment of the 1973 embargo at least seventeen federal agencies were involved in compiling and reporting information, most of it obtained from industry on a voluntary basis.[27] The Bureau of Mines in the Department of Interior, for instance, obtained much of its data from the annual reports of the American Petroleum Institute and the American Gas Association. This voluntary information was difficult to check because industry feared the release of proprietary information, and so what information was published tended to be very highly aggregated.[28]

The government's problem in gathering information about energy matters was presaged in 1962 in a report by the Kennedy administration's Petroleum Study Committee. The report argued that "satisfactory information concerning petroleum reserves, productive capacity, and deliverability, and their expansibility under normal conditions is seriously lacking. Suitable cost information is even more seriously lacking. A great deal of fragmentary and sometimes contradictory data is available." The report concluded that "corresponding data and analytical shortcomings are to be found in regard to the interrelationships of different segments of the economy. There is, in addition, a related inadequacy in analytic studies. These undesirably limit the conclusiveness of any petroleum study under existing circumstances."[29]

[26]Bruce F. Freed, "Energy Report/Government Seeks Ways to Verify Energy Data," *National Journal Reports,* February 2, 1974, p. 278.

[27]The problem of government dependence on industry information runs through many congressional hearings on energy. In testimony before the Church Subcommittee on Multinational Corporations and American Foreign Policy, James E. Akins, from the Office of Fuels and Energy, Department of State, prefaced his remarks on U.S. oil company negotiations with Libya: "We were relying on the companies for information and we still are. We possibly are going to be in a position sometime to require them to give us information but we have not been in that position in the past and we are not there now and I can be excused from talking about specific company positions inside these negotiations." Senator Church asked: "Why is it that the U.S. Government is in this position with respect to the oil companies? Why can't we obtain information as a matter of right?" Atkins retorted: "Well, I think is it probably the way the Government works." "Multinational Petroleum Companies and Foreign Policy," Hearings before the Subcommittee on Multinational Corporations, Committee on Foreign Relations, U.S. Senate, 93d Cong., 1st sess., Vol. 5, October 11, 1973, p. 9.

[28]See Freed, "Energy Report," February 2, 1974, p. 279, and more generally Neil De Marchi, "The Ford Administration: Energy as a Political Good," in Goodwin, *Energy in Perspective,* p. 292.

[29]Quoted in Bruce F. Freed, "Energy Report/Government Seeks Ways to Verify Energy Data," *National Journal Reports,* February 23, 1974, p. 282.

Twelve years later another major report echoed these findings. The General Accounting Office prepared a report, published in February 1974, for Senator Henry Jackson's Interior Committee. The study listed such problems in the energy area as the voluntary reporting of industry information, which does not "provide the federal government with the assurance that needed data will be available," and questions about its credibility. The report noted that the "federal government has been unable to demonstrate convincingly the nature and extent of the energy shortage, in large measure because of the lack of independently developed or independently verified data." In addition, the adequacy of information also caused problems. The absence of basic data on petroleum inventories, as well as energy distribution and consumption patterns, impeded government planning. Finally, until the establishment of the Office of Energy Data and Analysis in the Interior Department, which was later moved to the Federal Energy Office, there existed no agency empowered to analyze data on a continuous basis. "Perhaps the most crucial need," the report concluded, "is for analyses of energy data from the perspective of identified energy problems, rather than from the vastly different perspective of individual agencies and programs."[30]

The administrative response to this information problem led directly to the data modeling and analysis of Project Independence. Project officials themselves aimed to bring authoritative information to the center of policy deliberations; as I have suggested, this enterprise was ultimately unsuccessful. Project officials were not well received within the executive bureaucracy, nor did they have regular, direct access to the president. Congressional politicians, wary of both executive and industry data and analysis, also sought to establish a source of authoritative information, but unlike that of Federal Energy Administration officials, congressional sentiment favored an independent agency sheltered from policy planning. This idea was embodied in a bill proposed by Senators Gaylord Nelson and Henry Jackson in 1974.

The bill's approach was not welcomed by the Federal Energy Office (or later by the FEA). FEO assistant administrator Eric Zausner, in the midst of Project Independence preparations, argued: "We don't want a separate information agency established. We need a lot of information for the operation of our programs and we want to collect it

[30]"Energy Information Needs—Study by the General Accounting Office," Prepared at the request of the Committee on Interior and Insular Affairs, U.S. Senate, February 6, 1974, pp. 4–6.

ourselves. We have the expertise to collect the information and we know what to collect and assess."[31]

Project Independence sought to provide an authoritative body of experts and analysis, but it failed. In its place an independent body, separated from planning and policy deliberations, eventually became the government's response to information shortcomings. The organization of government thus adapted to the call for new sources and forms of information and expertise, but it adapted in a way that left the specialists outside the core of the policy process.

ENERGY INDEPENDENCE AUTHORITY

Nixon's energy independence proposal met with skepticism almost at the moment of its unveiling.[32] In many respects the articulation of the goal was designed to support international efforts by the administration to get industrial consumer nations to cooperate in undercutting OPEC's oil-pricing power. It also reflected the attempt of a president, under seige by the Watergate investigation, to rally the public and Congress to a major national cause. The project did, however, produce an ambitious modeling effort to develop the policy sophistication necessary to deploy state resources and stimulate domestic production. Yet the meager payoff revealed not only the limitations of policy competence within the administration but also the limitations on the government's ability to gain that competence.

Nixon's officials soon began to concede that the national economy would remain dependent on foreign sources of energy for some time to come. Energy research and development got under way as the newly established Energy Research and Development Administration (ERDA) geared up to propose projects and spend money on promising energy technologies for the long term. However, support persisted within the Nixon administration, and in the Ford administration to follow, for direct government sponsorship to stimulate the immediate production of domestic energy.

The Rockefeller Proposal

Within the Ford administration the most ambitious proposal for government involvement in energy production came from Vice-Presi-

[31]Quoted in Freed, "Energy Report," February 23, 1974, p. 282.
[32]E.g., Ernest Holsendorph, "Oil Independence—It Seems Unlikely," *New York Times*, December 2, 1973, sec. II, pp. 1, 15.

dent Nelson Rockefeller. In the summer of 1975 Rockefeller proposed within administration councils a $100 billion Energy Resources
Finance Corporation. This corporation would signal a major departure for the public role in the energy sector. At the heart of Rockefeller's proposal was a new commitment by the federal government to
underwrite, through loans and loan guarantees, energy projects that
the market would not sustain on its own. As Rockefeller's staff prepared the proposal, the vice-president began asking administration
officials to identify energy projects that were being held back in the
private sector because of financial problems.[33]

Although considerable opposition developed within the administration, Ford eventually agreed to the proposal. In a speech on
September 22, 1975, to an AFL-CIO convention, Ford said: "You and
I know we can produce our own energy. You and I know we can
protect ourselves against increases in price by foreign nations. You
and I know we can provide more jobs. And you and I know we can
bring an end to the intolerable situation in which America exports
more than $25 billion annually to pay for imported oil while plenty of
energy is potentially available right here at home." Noting that Congress had failed to act on earlier energy proposals, Ford proposed the
$100 billion government corporation. The corporation would be em-
powered to "take any appropriate financial action" needed to stimulate energy production projects that could not be financed under
prevailing circumstances. The corporation, Ford said, would engage
in the "crash development" of domestic energy resources, including
"new technologies to support or directly produce or transport American energy; technologies to support American nuclear development;
and electrical power from American coal, nuclear, and geothermal
sources."[34]

Ford sent the Energy Independence Authority Act to the Congress
on October 10, 1975. In submitting the act, Ford noted: "It is estimated that the capital requirements for energy independence will
total about $600 billion over the next ten years. Risks are such in
many of the projects necessary to develop domestic energy resources
and reduce consumption that private capital markets will not provide
necessary financing. The uncertainties associated with new technologies inhibit the flow of capital." The purpose of the authority was to

[33]De Marchi, "The Ford Administration," p. 518.
[34]Gerald Ford, "Remarks in San Francisco at the Annual Convention of the AFL-
CIO Building and Construction Trade Department," September 22, 1975, in *Public
Papers of the Presidents of the United States, Gerald Ford, 1975*, vol. II (Washington: GPO,
1977), pp. 1493, 1495-96.

assure the flow of capital into domestic energy production and to provide financial assistance where private-sector financing was inadequate. To this end the Energy Independence Authority (EIA) was to be constituted as a corporate body and, therefore, function outside established administrative bodies. To ensure autonomy and political neutrality, the authority was to be guided by a presidentially appointed and congressionally approved five-member board of directors, composed of representatives from both national parties.[35]

The authority was to be self-liquidating, mandated to terminate its investing operations after seven years and its institutional operations after ten years. During this period the authority was to be authorized to provide $100 billion in financial assistance to approved energy projects. Assistance could take the form of loans, loan or price guarantees, purchase of equity securities, or construction of facilities for lease purposes. Funds would be raised by the sale of up to $25 billion in equity securities and the issuance of up to $75 billion in government guaranteed obligations. In addition to stimulating investment within the energy industry, the authority was to promote regulatory reform within government—it was empowered to simplify and coordinate federal licensing and regulatory decisions that affected energy development. So constituted, the new government corporate body was to act at the intersection of government and business, facilitating and encouraging investment in domestic energy production.

Hearings were held on the EIA Act in April 1976. Vice-President Rockefeller argued that OPEC price increases had been manifestly political and that the American response would have to take a similar form. Domestic energy production needed to be stimulated beyond what established market processes would sustain. Thus the proposed EIA should be judged, he argued, for its political intentions and cast as a response to the new realities in international energy. "We've got a situation in the Middle East right now," Rockefeller argued, "that could blow up tomorrow and we could have another war. We could have another boycott. The east coast is now dependent 75 percent on energy from abroad. In 2 years we will be importing 25 percent of our energy from Arab countries because it's low-sulfur oil. . . If that's cut off, we're going to have absolute economic and social chaos on the east coast because you can't transport oil to the east coast from other parts of the country."[36] The EIA was a response not unlike earlier

[35]U.S. Congress, Documents and Report, *House Documents,* 94th Cong., 1st sess., vol. 4 (1975), pp. 1, 13–14.

[36]"Energy Independence Authority Act of 1975," Hearings before the Committee on Banking, Housing, and Urban Affairs, U.S. Senate, 94th Cong., 2d sess., April 12, 1976, pp. 2–9.

government responses to crisis. As examples he cited the Hoover administration's Reconstruction Finance Corporation and American government financing of aluminum and rubber projects during World War II.

The Energy Independence Authority proposal gathered support from policy specialists interested in confronting OPEC directly. Walt Rostow, a former foreign policy official, for example, argued that "until something like [the] Authority is at work and the rate of American investment in energy and energy conservation is rising rapidly, I doubt that a fruitful negotiation between OPEC and the major oil importers will be possible." The act was also supported by administration officials who saw it as necessary because Congress continued to insist on price controls. The message these officials gave was that if Congress failed to act on modest proposals for price decontrol and taxing, stronger measures, such as those embodied in the EIA, would be necessary. FEA chief Frank Zarb, for example, reminded critics that it was Congress which had failed to follow Ford's 1975 plan to decontrol prices for domestic crude oil immediately. But he went on to claim that even with decontrol, "we have lost so much time because of our sell-out to cheap oil throughout the 1960s and early 1970s and [because we have] neglected so much our domestic technology and capability in the coal area and the nuclear area."[37]

Senator Jackson's Plan

The Rockefeller proposal contained ideas that had earlier surfaced in Congress. Senator Henry Jackson, chairman of the Senate Interior and Insular Affairs Committee, had already proposed and held hearings on legislation to create a National Energy Production Board (NEPB). His bill also envisaged a rapid and massive mobilization of domestic energy production. The NEPB would be less concerned with capital investment than with the immediate development of energy resources on public lands. It was to centalize government regulatory and licensing procedures in "the public and private sectors for increasing the exploration, development, production, and transportation of domestic coal, oil, and natural gas."[38] The board was to be a central body coordinating government policy; it would propose additional legislation; and, with a trust fund with an annual appropriation

[37]Walt W. Rostow, "Energy Independence Authority Act of 1975," Hearings before the Committee on Banking, Housing, and Urban Affairs, U.S. Senate, 94th Cong., 2d sess., April 12, 1976, p. 88.
[38]"National Energy Production Board," Hearings before the Committee on Interior and Insular Affairs, U.S. Senate, 94th Cong., 1st sess., March 20, 1975, p. 28.

of $1 billion, it would invest in public and private energy projects.

Support for the board came from senators and other officials who conceived of the energy challenge in terms of national security.[39] Testifying in favor of the Jackson program, for example, was former treasury secretary Henry Fowler, who likened the current energy crisis to earlier national energy emergencies. "There are some lessons of experience relevant to the energy problem to be learned," Fowler argued, "from the two most recent historical efforts to organize massive national production and supply programs in conjunction with World War II and the so-called Korean war mobilization." He went on to note the experiences of the War Production Board and the Petroleum Administration for War of the World War II and the Defense Production Act of 1950.[40]

Support for both the Rockefeller and the Jackson programs tended to come from the same government and private quarters. The investment banker Felix G. Rohatyn, for example, testified in favor of the Jackson proposal but stressed the need for an Energy Finance Corporation styled after the Hoover administration's Reconstruction Finance Corporation. In all cases supporters looked to earlier historical experience. Noting that "history can be instructive," Senator Jackson argued that "we need a single, specialized, mission-oriented action agency empowered to prepare and overcome bottlenecks, and marshal Governmentwide resources for a major domestic energy production program."[41]

Opposition to the Program

The Rockefeller proposal, unlike the Jackson plan, had received support from President Ford. If the government was going to estab-

[39]Jackson's proposal was cosponsored by Senators Magnuson, Bayh, Leahy, and Stevenson.

[40]"National Energy Production Board," Hearings, March 20, 1975, p. 49. Fowler's argument reflected the conclusions of a private study he had chaired for the Atlantic Council of the United States, a private study group. The report argued: "The Administration and Congress should arrange urgently to concentrate in a single all powerful agency reporting directly to the President, all of the government's authority to act on production of energy, whether through changes in the economic and legal climate to accelerate investment, expedited regulatory decisionmaking, or direct action by government to assure the availability of enough energy to satisfy our national needs and our national interest in international energy cooperation and avoid inadequacy of energy supply whether threatened by the pluralistic processes of the marketplace, divided authority in public regulation, or action by the producer's cartel." Atlantic Council of the United States, "World Energy and U.S. Leadership," *Policy Papers,* January 1975.

[41]Hearings, "National Energy Production Board"; for Rohatyn, March 20, 1975, p. 131; for Jackson, April 14, 1975, pp. 141–42.

lish a new, centralized authority for energy production, the EIA was the most likely vehicle. Support for the Energy Independence Authority, however, did not run deep within the Ford administration. Even in the spring and summer of 1975, as the Rockefeller proposal was being formulated, opposition surfaced among administration officials. It centered within the Council of Economic Advisors. Council officials circulated memoranda in administration circles questioning the need for the government to encourage investment in energy production. CEA chairman Alan Greenspan and Treasury Secretary William Simon were vocal critics of the proposal. Energy officials in the Federal Energy Administration also opposed the plan, more for reasons of bureaucratic control. The EIA, it was reported, might pose a bureaucratic threat to their operations. Finally, opposition to the Rockefeller plan came from the director of the Office of Management and Budget, James T. Lynn, who argued that Congress could manipulate the Energy Resource Finance Corporation to extend its life and channel resources into favored energy projects.[42]

In Congress, opposition came from politicans of different political persuasions. Some Democrats had supported a new energy finance authority—indeed, a Democratic energy policy alternative to Ford's January 1975 plan (which did not then include the Rockefeller proposal) had as its centerpiece Senator Jackson's National Energy Production Board.[43] Nonetheless, congressional liberals were suspicious of the proposed corporation's ties to the big energy producers. Representative Henry Reuss, chairman of the House Banking, Currency, and Housing Committee, which had jurisdiction over the proposal, said the corporation would be "grossly inflated, fiscally irresponsible and susceptible to political manipulation."[44]

Prevailing sentiment among liberal Democrats was expressed by Senator Edward Kennedy, who argued that the finance corporation would drain capital from other social projects and subsidize the big oil corporations. The program, he said, is "likely to create a sharp drain on capital toward energy and away from other sectors of the economy." In addition, it "will be using public money to subsidize energy development projects. In turn, these projects are likely to bring high profits to private energy corporations—almost inevitably the major oil companies under the administration plan—without any evidence

[42]De Marchi, "The Ford Administration," pp. 518–20. On FEA opposition see David S. Broder, "Ford Asks Creation of Energy Corporation," *Washington Post*, September 23, 1975, p. A11.

[43]*The Congressional Program of Economic Recovery and Energy Sufficiency* (Washington: GPO, 1975); De Marchi, "The Ford Administration," p. 491.

[44]"$100 Billion Energy Agency Formally Proposed," *Congressional Quarterly*, October 18, 1975, p. 2238.

that these rich profits can be justified by benefits in other ways to the American economy and the American people."[45]

The coalition of opponents also included environmentalists and conservatives from oil-producing states, who had already opposed the Jackson plan.[46] Environmentalists were concerned with the prospects of added air and water pollution from large-scale coal and shale-oil plants. Conservatives saw in the corporation a competitive threat to existing energy producers.[47]

The Rockefeller proposal eventually failed because it did not have a well-established constituency to build congressional and executive support. Its proponents within the Ford administration were few and not well-positioned. One analyst has noted that the plan lacked what he calls an "operational and effective champion in the mainstream of the policy process."[48] Moreover, the rationale for the program was vulnerable to attack from both liberals and conservatives. Many liberals doubted the EIA would be anything more than a massive subsidy program for the energy industry. Some of these same politicians questioned the assumption that energy demand could not be substantially reduced through conservation. For conservatives, the threat was to the primacy of a private system of energy production. The national crisis of energy supply, which by 1975 had greatly diminished, was not sufficient to override this crosscutting opposition.

In February 1976 President Ford renewed his energy requests. Little of the earlier Ford proposals had passed Congress. Among the proposals the administration urged Congress to act upon was the government finance corporation. But with opposition inside the administration and vocal critics in Congress, the renewed request was no more than a formal statement of continued support. Little on the political horizon suggested that favorable congressional action would be forthcoming.[49]

THE ENERGY CORPORATION OF AMERICA

One far-reaching proposal for institutional change came from the Senate. Senator Adlai Stevenson III led an effort to establish a gov-

[45]*Congressional Record*, Senate, October 8, 1975, pp. 32175–77.

[46]See comments by Senators Packwood, Proxmire, and Metzenbaum in "Energy Independence Authority Act of 1975," Hearings before the Banking, Housing, and Urban Affairs Committee, U.S. Senate, 94th Cong., 2d sess., April 12, 1976.

[47]David Burnham, "Environmentalists and Oil-State Conservatives Likely to Fight Plan for Energy Agency," *New York Times*, September 23, 1975, p. 15.

[48]Mel Horwitch, "The Convergence Factor for Large-Scale Programs: The American Synfuels Experience as a Case in Point," MIT Energy Laboratory *Working Paper*, December 8, 1982, p. 23.

[49]*Congressional Quarterly*, March 6, 1976, p. 525.

ernment energy corporation that would engage in the purchase, exploration, and production of petroleum outside American borders. It would also have exclusive rights to develop resources on government-owned land, and it would take over planning functions from the Department of Energy. Thus it would be involved in formulating government plans to stimulate the development of alternative energy technologies. Congressional debate on this proposal came as early as December 1973 (a more elaborate proposal was made in 1979). The rationale for the energy corporation and the opposition to it both exposed the limits to government involvement in the energy area even at moments of crisis.

Rationale and Proposals

The rationale for a government energy corporation in the United States diverged little from what lay behind similar initiatives in Europe. First, the Stevenson bill sought to provide a government instrument to redirect production away from the Middle East. Whereas multinational oil production and marketing was driven by established commercial contracts and relationships, a government firm would attempt to alter these patterns. Second, a government corporation would be positioned to negotiate broader deals with producing nations for secure and long-term contracts. A major oil company primarily has narrowly commercial bargaining resources, but a government can bring technological, financial, and political resources to a deal. Finally, a government corporation would be able to develop the analytical expertise and information-gathering capacities needed to monitor change in the energy sector. The corporation would provide a "wedge" into the sector and thus mitigate the prevailing dependence of government on private companies for information on which to base decisions.[50]

In its first formulation, in 1973, the proposal took the shape of an amendment to a larger energy package, the Consumer Energy Act. One of Stevenson's cosponsors, Senator Walter Mondale, gave an extended rationale for the corporation.[51] First, the government corporation would aggressively explore and develop resources on federal lands. Second, the corporation would supply energy information to government. "We are forced to rely on industry statistics which may or may not be accurate," Mondale said. "This corporation would go a long way toward giving those of us in Government a much better

---

[50]Interview, Senator Adlai Stevenson, III, June 13, 1984.

[51]*Congressional Quarterly*, January 26, 1974, pp. 174–75. The cosponsors were Senators Abourezk, Philip Hart, McGovern, McIntyre, Metcalf, Mondale, and Moss.

idea of the real cost of exploring and developing oil and natural gas resources." Third, the corporation would sell its oil and natural gas resources in such a way as to promote competition within the energy industry. If it did market oil and natural gas, the corporation would be required to give preference to independent firms and to governments. Fourth, the corporation would be able to draw on the federal treasury for up to $50 million a year over ten years. But, Mondale argued, the corporation would likely realize a profit on the sale of resources from federal lands, profits that would go back into the national treasury. Finally, Mondale claimed, the corporation would "help restore competition to an industry badly in need of a stiff dose of the competitive spirit."[52]

Although hearings were held on the Stevenson proposal, no committee action was taken on the amendment. The idea did not have enough support, and the plan would have to wait until developments provided a new opportunity for its debate.[53]

In 1979 Stevenson again brought the proposal to the attention of his congressional colleagues. President Jimmy Carter had left the annual economic summit of the industrial countries, in Toyko, with a pledge to reduce American oil imports. A second dramatic rise in the price of oil was just then shaking the industrial countries. The experience was making it clear to politicians and others that the United States had not adequately reduced its dependence on imported oil.

Hearings on the proposal, renamed the Energy Corporation of America, were held in July 1979 before the House Ways and Means Committee. A companion bill, the Energy Bank of America, was introduced in the Senate. Ways and Means entertained a variety of bills to create a government authority to regulate oil imports. Those representatives who proposed some form of a government importing company uniformly noted that the United States was the only major industrial country without a government apparatus to control oil imports. The sentiment for change was expressed by Representative Benjamin S. Rosenthal of New York: "If ever there was a time or a need for such a Federal responsibility via a Federal corporation, the time is now. If the President is going to lead the Nation toward a restriction on foreign oil imports, then the present Jerry-built system that has existed for 60 to 70 years permitting the major multinational

---

[52]"Consumer Energy Act of 1974," Hearings before the Committee on Commerce, U.S. Senate, 93d Cong., 1st and 2d sess., December 12, 1973, pp. 1000, 1001.

[53]Despite support from committee chairman Magnuson, the Consumer Energy Act did not gain a majority vote within the Commerce Committee. The bill was reintroduced in 1975. *Congressional Quarterly*, February 22, 1975, p. 364.

oil companies to control the destiny of this country has to come to an abrupt, if not screeching, halt."[54]

Proposals varied in the degree to which the government import authority would resemble a corporate entity. Several congressional representatives advocated an importing authority that would auction off a set quota of oil imports to domestic refiners and marketers. Representative Paul Findley from Ohio proposed an auction authority. "Public auction sales," he argued, "would generate reliable public data on prices and volumes, both before and after refining. Such information would make it relatively easy to detect attempts by large oil companies to take advantage of any monopoly power."[55]

Stevenson's proposal envisaged an extensive mandate for a government energy corporation. His corporation would be not just an import authority but also authorized actually to engage in energy exploration and development both at home and abroad. It would even be able to consummate direct bilateral deals with producing countries.

Opposition to the Program

There was not a great deal of support for the energy corporation idea in either the Congress or the executive branch. Senator Jackson supported Stevenson, as did Senator George McGovern, and other liberal senators cosponsored the proposal. In the 1979 hearings, several House members testified in favor of the bill, and other members introduced legislation to provide for a similar government corporation.

Stevenson gives two major reasons for the failure of the ECA proposal. The first was the vigorous opposition of major and independent oil companies; their active attacks on the proposal, Stevenson later argued, were critical to the absence of congressional support. The bill attempted to split the oil industry and bring independent producers and marketers in line behind the proposal. To this end the ECA was mandated to develop American public lands and give favored access to small, independent oil refiners and marketers. This tactic was not successful. The anticipated government threat to industry operations was a stronger consideration than the promise of favored access, and the energy industry tended to stand together in opposition.

Second, Stevenson argues, the ideological attack of various oppo-

---

[54]"Oil Import Policy Issues," Hearings before the Subcommittee on Trade on the Committee on Ways and Means, U.S. House, 96th Cong., 1st sess., July 16, 1979. p. 50.
[55]Ibid., July 17, 1979, p. 69.

nents weighed heavily in the final outcome. The idea that this new government involvement transgressed traditional lines of public and private responsibility, that it was "radical". and "socialist" in inspiration, prevented active support from congressional members in spite of the suspicion among Democratic congressmen that the majors were operating in ways inimical to the national interest.[56]

In effect, the Stevenson proposal, and others like it, were attempting to transform the traditional relationship between government and industry. The government corporation would bring the state into the sector as a player, allowing the government to develop a competence to evaluate price and production patterns on a more equal basis with industry. At the same time the proposal envisaged this shift as resulting in less direct regulation of the oil and natural gas industry. Thus Stevenson and others anticipated that the government entity would actually promote competition. Furthermore, they understood their program as drawing upon earlier and successful government efforts in resource development. Experience with the Tennessee Valley Authority provided a historical precedent and a rationale for the new public corporation. Despite this rationale, however, the proposal was a departure that could not gather together a successful coalition of supporters.

CARTER'S ENERGY SECURITY ACT

In 1979 the U.S. economy was actually more dependent on foreign imported oil than it had been in October 1973. Much of President Carter's energy legislation had not passed Congress. The Iranian revolution, which would trigger another dramatic sequence of OPEC oil price rises, was still several months away, but many analysts could hear the ticking of an oil price "time bomb."[57] On July 15, 1979, Carter delivered a national speech on his presidency. Part of that speech focused on energy, and a centerpiece of his new energy program was an ambitious proposal to develop American energy resources. He stated: "I propose the creation of an energy security corporation to lead this effort to replace 2.5 million barrels of imported oil per day by 1990. The corporation will issue up to $5 billion in

[56]Interview, June 13, 1984.
[57]Richard Corrigan, "The Oil Price Revolution Is Alive," *National Journal*, March 3, 1979, p. 360; Robert J. Samuelson, "The Oil Price Time Bomb," *National Journal* 11 (June 2, 1979), 916.

energy bonds—in small denominations so that average Americans can invest directly in America's energy security."[58] A week later Carter sent a message to Congress outlining his proposed Energy Security Corporation (ESC).

The Carter proposal was an attempt to bring the federal government into the development and commercialization of alternative energy technologies. After the political struggles finished, what the Carter administration actually got was a program to begin a difficult, and largely unsuccessful, process of encouraging commercial projects to develop synthetic fuels. The ambitiousness of the proposal and the modesty of the outcome tell us much about the limits on an American government presence in the energy sector.

Historical Antecedents

Proposals for a government synfuels program had circulated within Congress and the executive branch since the end of World War II. In 1948, in the midst of a scare over the availability of fuel, the secretaries of defense and interior proposed an $8–9 billion synthetic fuels program. Earlier research had been stimulated in 1944 when the Congress passed the Synthetic Liquid Fuels Act, appropriating $85 million for research and the construction of demonstration facilities. The law had authorized the Bureau of Mines and the Department of Agriculture to develop new technologies to produce gas from coal, oil from shale, and a variety of liquid fuels from coal, lignite, and agricultural wastes. These early initiatives were small and moved rather slowly, however, because of opposition from the oil industry and subsequent controversies over cost estimates and technologies.[59] When the program was terminated in 1955, approximately $82 million had been spent on research and development.

Early proposals for synfuel subsidies were vigorously opposed by the petroleum industry. Between 1949 and 1953 bills to encourage synfuel projects were introduced in every session of Congress, backed by the White House, the Defense Department, the Council of Economic Advisers, and other key administrative bodies. However, a

---

[58]President Jimmy Carter, "Energy and National Goals: Address to the Nation, July 15, 1979," in *Weekly Compilation of Presidential Documents* 15, no. 29 (Washington, D.C.: Office of the Federal Register, July 23, 1979), p. 1239.

[59]Craufurd Goodwin, "The Truman Administration: Toward a National Energy Policy," in Goodwin, *Energy Policy in Perspective;* Richard E. Vietor, "The Synthetic Liquid Fuels Program: Energy Politics in the Truman Era," *Business History Review* 54 (Spring 1980), 1–34.

coalition of conservative Republicans and Southern Democrats successfully blocked the legislation.[60]

Much of the debate hinged on the comparative costs of oil and synfuels. The Bureau of Mines estimated in 1949 that gasoline synthesized from coal would cost 10 cents per gallon and from shale, 11.5 cents per gallon—virtually competitive with the prevailing costs of petroleum production. In 1951 the National Petroleum Council, the chief representative organization for the oil industry, estimated the costs of gasoline synthesized from coal as approaching 41.4 cents per gallon, and 16.2 cents per gallon for shale oil. A subsequent independent analysis, commissioned by the Department of Interior, estimated costs as somewhere between government and industry projections. This study also concluded that industry itself was not equipped to finance the start-up of a synfuels industry.[61] Both sides claimed victory; the dispute escalated. On one side was the Bureau of Mines, which had more than half of its budget committed to synfuels development. Allied with the bureau were the coal industry and agencies and officials within the national security community. Opposing was the petroleum industry, and giving voice to its position the National Petroleum Council.

The dispute could not be settled on simple analytic grounds. Studies and estimates varied too widely. Indeed, as one analyst concludes, "what the situation lacked was a strong and effective third party able to judge among the competing claims and to make firm policy recommendations with the broad public interest at heart."[62] The dispute over costs continued into 1952, and this delay was enough to kill the program.

The fate of these early proposals suggests that their appeal peaked at the moment of crisis. Because of the massive commitment of resources involved, a synfuels program would be viable only if couched in national security terms. But the energy payoff would most likely come after the crisis had abated. By 1952, for instance, the perceived crisis over fuel supply brought on by the Korean War had diminished. By that time the national security imperative was no longer clearly sufficient, and support for the initiative could not justify further struggles with the petroleum industry.

[60]Vietor, "Synthetic Liquid Fuels Program," pp. 14–15.

[61]Goodwin, "Truman Energy Policies toward Particular Energy Sources," in Goodwin, *Energy Policy in Perspective*, pp. 161–67; Horwitch, "The Convergence Factor," p. 15.

[62]Craufurd D. Goodwin, "Truman Administration Policies toward Particular Energy Sources," in Goodwin, *Energy Policy in Perspective*, p. 159.

Interest in synfuels reemerged in the 1960s, but not until 1971 was major new legislation proposed. Jennings Randolph, who as a congressman had been instrumental in the passage of the Synthetic Liquid Fuels Act of 1944, joined Henry Jackson in the Senate in sponsoring legislation to create a federal corporation for coal gasification. This proposal led to the inclusion of synthetic fuels in the federal Nonnuclear Energy Research and Development Act of 1974, which contained most of the principles and incentives for synfuel development that would reappear in Carter's Energy Security Act of 1980.[63]

The Nonnuclear Act of 1974 authorized the creation of the Energy Research and Development Administration (ERDA), giving it the use of a variety of assistance mechanisms, including joint federal–private sector corporations, cooperative agreements, purchase agreements, price supports, direct loans, and grants. Such mechanisms had become standard tools of federal support in a wide range of public policy areas. Agricultural policy relied on price supports; nuclear power policy was based on cooperative agreements; defense contractors relied on purchase agreements; scientific research was supported by federal grants. The Nonnuclear Act authorized the use of these typical instruments to promote new energy technologies.[64] Yet Congress did not authorize specific programs. It left those decisions to subsequent congressional budget allocations. Thus while Congress agreed in principle to promote the commercialization of new energy technologies, including synfuels, it postponed the legislative battle over synfuels.

The Ford administration endorsed the idea of a synthetic fuels program in the president's 1975 State of the Union message. Ford urged the creation of a program to produce the equivalent of one million barrels of oil per day by 1985. Rockefeller's Energy Resource Finance Corporation, which Ford publically proposed in October 1975, also would have been actively engaged in promoting synfuel projects.[65] The administration anticipated folding the synfuels programs into the Energy Independence Authority once that larger administrative body was established.[66] By September 1975, however, administration hopes for a massive synfuels industry had diminished.

[63]Carroll E. Watts, "The U.S. Synthetic Fuels Program: An Overview," Government Research Corporation, May 4, 1981, p. 1.

[64]Ibid., pp. 2–3.

[65]See "Energy Independence Authority Act of 1975," Hearings, Committee on Banking, Housing and Urban Affairs, U.S. Senate, 94th Cong., 2d sess., April and May 1976.

[66]"Congress Debates Synthetic Fuel Proposals," *Congressional Quarterly*, November 8, 1975, p. 2398.

An interagency report prepared by the Natural Resources Council acknowledged that a series of problems stood in the way of large-scale production; chief among them were environmental problems and a shortage of water in key regions. Nonetheless, the report reaffirmed the desirability of establishing a vigorous, if reduced, program of synfuel production.[67]

The immediate obstacle to passage of a program was resistance within Congress, particularly in the House. During consideration of the Nonnuclear Act, the Senate had actually supported the idea of a joint public-private corporation to develop synthetic energy for ten years. It also supported funds for ERDA in the FY 1976 authorization bill, to promote synfuels commercialization. However, both measures were defeated in the House. Liberal House Democrats were concerned that the program was a subsidy for big business. "When we talk about loan guarantees . . . we are talking about subsidizing commercial ventures with the taxpayer's money," one congressman commented during hearings in October 1975. Others focused on the environmental and social impacts of a massive synfuels program. Representative Ken Hechler, chairman of the fossil fuels subcommittee, noted: "The entire emphasis of this program has been that industry has risks. These risks must be minimized by loan guarantees. The contention of this committee is that people have risks, communities have risks, taxpayers have risks. . . These . . . must be considered at the same time as the other legislation."[68]

In late 1977, while the Congress was debating the legislation that would create the Department of Energy, the House agreed in principle to loan guarantees for a synfuels program, but the bill contained no funding provision.[69] Later, the Department of Energy would request funds for synfuel commercialization projects, but at that juncture the Carter administration itself rejected the proposal. As late as 1979, support for a commercial synfuels industry could not be pieced together.

The major obstacle to legislation in the House was an unlikely coalition of liberal and conservative congressmen. Liberals favored conservation, smaller-scale technologies, and less environmentally disruptive approaches to energy policy. They were also skeptical about subsidizing projects that would likely remain in the hands of big ener-

[67]Thomas O'Toole, "U.S. Reduces 1985 Goal for Oil Substitute," *Washington Post*, September 21, 1975, pp. 1, 8.
[68]Both quoted in "Congress Debates Synthetic Fuels Proposals," *Congressional Quarterly*, November 8, 1975, p. 2398.
[69]*Department of Energy Act of 1978, Civilian Applications*, Public Law 95-238.

gy companies. Conservatives opposed further interventions in energy markets, and they did not want any enlargement of the executive bureaucracy. As it had with the Rockefeller proposal, this coalition successfully blocked legislation for a synfuels program. Not until 1979, as a second oil shortage and massive OPEC price rise created a new political call for action, did the Carter administration and liberal House members move to support legislation.

The Carter Plan

Although Carter embraced and promoted the synfuel idea in 1979, the shape of the program and its modest success depended as much upon initiatives taken in Congress as upon the president. Within the House, Representative William Moorehead was instrumental in crafting a proposal that could eventually gain House passage. In his initial proposal in 1976, Moorehead shifted the rationale for a synfuels program from domestic energy policy to national security, styling his proposal along the lines of the 1950 Defense Production Act. That act had authorized the federal government to mobilize industry for national defense purposes. Moorehead drew up legislation that amended the Defense Production Act, authorizing the government to encourage the development of synfuels.

Moorehead was thus able to draw on support from the Defense Department. In presenting his proposal to the Economic Stabilization Subcommittee of the House Banking Committee, in 1979, he received favorable testimony from Defense officials. The undersecretary of defense for research and engineering, Ruth Davis, provided a defense rationale for the program: "the continued dependence of our forces on liquid hydrocarbon fuels, when combined with the Nation's growing dependence on vulnerable sources of foreign oil supplies, now approaching 50 percent of total U.S. requirements, gives cause for grave concern about our ability to adequately provide for national security. We must find energy alternatives . . . alternatives that are insensitive to capricious economic and geopolitical actions."[70]

In addition to changing the political debate with a national security rationale, Moorehead was able to move his legislation through a less hostile committee. The bill was channeled through the House Banking Committee rather than the House Interstate and Foreign Com-

[70]Ruth Davis, "To Extend and Amend the Defense Production Act of 1950," Hearings before the Subcommittee on Economic Stabilization of the Committee on Banking, Finance and Urban Affairs, 96th Cong., 1st Sess., March 13, 1979, p. 23.

merce Committee, which had defeated earlier synfuels proposals. Supported by House leaders, Moorehead had the bill move directly to the Rules Committee and to the House floor. On June 26, 1979, the House of Representatives passed his proposal by a vote of 368 to 25. Moorehead's tactical maneuvers were important, but the new energy emergency of 1979 also generated support previously lacking. The Iranian revolution had resulted in partial cutoff of Persian Gulf oil, and this shortfall led to gasoline shortages in the United States. Somewhat later on, OPEC again raised oil prices dramatically.

The Moorehead bill was a rather simple proposal, relying primarily on purchase agreements and price supports to encourage synfuel production. The government could buy the synthetic fuel and use it or resell it. If the government did not buy the fuel, it would pay the difference between an agreed upon price and the prevailing market price. The bill contained other mechanisms, among them loan guarantees, direct loans, and government-owned and contractor-operated production facilities. But these other tools generally required further congressional action.

The guaranteed price arrangement appealed to a variety of groups. The Committee for Economic Development (CED), a private business organization, in a study released in July 1979, argued that this method was the least obtrusive of the various government proposals. "It does not add to government interference with market prices and the extent of the subsidy, if any, is clearly apparent," the CED noted. "Also, the government does not get involved in the internal operations of energy facilities or in the actual design of the facility."[71] The Department of Defense, familiar with this form of cooperation, also supported the Moorehead approach.

The Senate was debating two proposals in early 1979. One was Senator Jackson's bill, which built on the Nonnuclear Act authority, with funding requests for synfuel projects channeled through the Department of Energy. The bill also contained provisions for a wide variety of energy technologies, such as solar, geothermal, and conservation. Senator Pete Domenici, of the Energy and Natural Resources Committee, proposed a different approach: to create a government corporation mandated to develop synfuels and other alternative energy technologies. The corporation would have $75 billion at its disposal.

At this point President Carter entered the debate. Support for synfuels legislation was already widespread in Congress, and Carter's

---

[71]Quoted in ibid., p. 10.

participation affected less the final passage of the legislation than its shape. On July 15, Carter publicly committed his administration to the proposal to found a synfuels corporation.[72]

In addition to the Energy Security Corporation, Carter proposed a new administrative board, the Energy Mobilization Board. This board, styled after wartime arrangements and situated within the Executive Office of the President, would be empowered to expedite energy production by breaking through regulatory tangles. On July 15, 1979, Carter announced: "I will urge Congress to create an energy mobilization board which, like the War Production Board in World War II, will have the responsibility and authority to cut through the redtape, the delays, and the endless roadblocks to completing key energy projects."[73] The board would have authority to overturn state and local land use, health, and safety laws. Procedural challenges to the board's decisions, furthermore, were to be limited to judicial review by federal courts of appeal. In a document prepared by the Justice Department's Office of Legal Council, John M. Harmon argued that the proposed federal preemptions of state and local authorities were permissible under the Commerce Clause of the Constitution, because the board's actions would promote "the important national interests of reducing oil imports and increasing domestic energy production."[74] The proposed board began almost immediately to run into opposition, however, particularly (and not surprisingly) among state and local officials.

The Energy Security Corporation would be authorized to borrow up to $88 billion from the Treasury between 1980 and 1990. It would, in Carter's words, "be an independent, government-sponsored enterprise with Federal charter." He went on to say:

> It will be located outside the executive branch, independent of any government agency. As such, the ESC should be able to staff, operate and take action unlike agencies of the executive branch. It should be able to act quickly and decisively consistent with its broad charter and goals. It must work with private industry and should not be hobbled by time-consuming and burdensome provisions of law which might increase the reluctance of business to become insnared with another Federal agency.

[72]See Richard Corrigan, "Congress Has Synfuels Fever," *National Journal,* June 23, 1979, p. 1050, and Dick Kirschten and Robert J. Samuelson, "Carter's Newest Energy Goals—Can We Get from Here to There?" *National Journal,* July 7, 1979, pp. 1192–1202.

[73]Jimmy Carter, "Energy and National Goals," July 15, 1979, *Presidential Documents,* p. 1240.

[74]Quoted in Dick Kirschten, "Cutting Energy Project Red Tape Raises Legal, Practical Questions," *National Journal,* September 9, 1979, p. 1449.

Accordingly, the Corporation would not be subject to those provisions of law which govern the administration and operation of government agencies and government employment.[75]

The Senate was receptive to the Carter proposal and incorporated its provisions into the Domenici plan. Yet even before debate began, a House appropriations subcommittee approved a $19 billion Energy Security Reserve in the Treasury Department. From this reserve, Congress appropriated just over $2 billion for the Department of Energy to begin start-up studies and programs for synfuel development.[76] The funding package, contained in the Interior Appropriations Act of Fiscal Year 1980, was a breakthrough: for the first time funds had been appropriated for a major synfuels effort. And it had happened even before Congress finished considering Carter's Energy Security Act.

Final congressional passage of the Energy Security Act required a number of compromises. The Senate proposal included most of Carter's requests. The program would take the form of a government corporation, initially receiving $20 billion to finance various projects. At its disposal would be authority to provide purchase agreements, price supports, loan guarantees, joint ventures, and, if necessary, government-owned and contractor-operated production facilities. Also, the corporation could come back to Congress for an additional $68 billion in funds.[77]

The House bill involved Representative Moorehead's more modest $3 billion program organized around Department of Defense purchase agreements. A new administrative organization would not be involved. Unlike the Senate bill, the House version gave the president wide scope for discretion. The compromise eventually achieved made the House bill, which amended the Defense Production Act of 1950, an interim measure while the Synfuels Corporation was organizing itself; thereafter the defense production authority would exist on a stand-by basis.

Carter signed this legislation on June 30, 1980, declaring that the new program "gives us the weapons to wage and win the energy war."[78] However, the companion bill, to create an Energy Mobilization Board, did not survive House vote and was sent back to the

[75]Carter, Message to Congress, July 20, 1979.
[76]Harry Perry and Hans H. Landsberg, "Factors in the Development of a Major U.S. Synthetic Fuels Industry," *Annual Review of Energy* 6 (1981), 237–38.
[77]Watts, "U.S. Synthetic Fuels Program," p. 12.
[78]*New York Times,* July 1, 1980, pp. D1, D4.

House-Senate conference committee. The Congress had created a new funding agency with powers to promote synfuel production, but it would not grant extraordinary powers to the president to bypass administrative and regulatory processes in the licensing and approval of energy projects. Carter's victory actually signaled the limits of executive discretion and authority. These mixed results bear further discussion.

Explaining the Outcome

The Synfuels Corporation was established, and by the end of the decade it embodied the only significant new role for government within the energy sector. Why did it succeed where other proposals failed? The straightforward explanation is that it did not step on important people's toes. The Synfuels Corporation was, in essence, a government spending program. It funded experimental and demonstration synfuels projects that would remain in private hands—in effect, it provided a financial subsidy to private business. This was by no means a radical departure in government-industry relations.

Second, the program was attractive to elements in Congress. Because it would support projects located in specific states, the program offered congressmen the chance to gather constitutent support. Unlike the other programs, it allowed for "credit claiming" by representatives.[79] In the implementation phase the program was particularly prone to efforts by legislators to service their local constituencies. One senator noted that during the budgeting phase, several congressmen actively sought and gained funding for synfuel projects that used such regional resources as shale, coal, and wood chips. And, he notes, this budgeting was conducted without even the appearance of staff analysis by which to rank promising fuel types and apportion resources accordingly.[80]

Beyond this, the Energy Security Corporation contained provisions with a wide appeal. The original Senate bill had eleven titles. The first created the synfuel corporation, the other ten provided incentives for alcohol fuels, biomass, geothermal energy, solar energy, and conservation. The bill was clearly designed to appeal to a coalition—it contained something for everyone.

Third, the Synfuels Corporation, at least at the level of program design, had been around for many years. It had support in some

[79]See David Mayhew, *Congress: The Electoral Connection* (New Haven: Yale University Press, 1974).
[80]Interview, Adlai Stevenson, III, June 13, 1984.

quarters of Congress that had been pushing the idea since World War II. A record of study and debate lay behind the idea, and unlike the other proposals examined in this chapter, this program had had time to take shape and develop support.

Finally, funding was made possible by the revenue to be extracted from the Windfall Profits Tax. The money could be put in the hands of an independent synfuels board, and so the executive branch would not be involved directly in energy production. The program was designed to allow a meshing with private industry.

At the same time the failure of Carter's request for a broad set of administative powers—the Energy Mobilization Board—exposed the limits of the delegation of power. Congress was hesitant to give the executive any authority that could not be circumscribed by annual budget reviews. And where new administrative authority was necessary, as in the synfuel corporation, Congress made sure that its corporate board would be independent of executive control.

CONCLUSION

To cope with problems of energy adjustment in the 1970s, U.S. government officials made various attempts to build additional state capacities. In effect, these proposals—ranging from the strengthening of the analytic competence of energy planners to direct government participation in energy finance and production—sought to redefine the state's role in the energy sector. They fit into a larger sequence of energy adjustment policies. Initial state-building proposals sought to increase the information and expertise on energy available to policy makers and to facilitate the ongoing international initiatives of the Nixon administration. As American diplomatic efforts were frustrated, executive and congressional officials advanced other proposals that envisaged even greater state involvement in the energy sector. These proposals, as we have seen, met with substantial opposition and (apart from the synfuels program) were easily defeated. Thus constrained, the search for a workable adjustment policy continued.

The defeat of these state-building initiatives has been explained in terms of the immediate opposition of various groups and individuals, but also in terms of the larger institutional setting in which policy struggles took place. Each proposal ran afoul of interest groups and public officials. Indeed, these state-building efforts generated a curious mixture of detractors. Vice-President Rockefeller's energy finance proposal and Senator Stevenson's government energy corporation each attracted the opposition of conservative and market-oriented officials both in Congress and in the executive branch. Yet many liberals were also troubled by their proposals. The emphasis on government involvement in a massive push to expand domestic production of energy attracted the opposition of environmentalists and those who wanted more attention to be paid to energy conservation. Other liberal Democrats opposed the state-building initiatives because they feared they were only new mechanisms by which government would subsidize the large energy firms.[81] A crossfire of opposition developed: conservatives feared the encroachment of state, liberals feared the new state capacities would be put to the wrong purposes or subverted by private interests.

The limited success of state-building efforts in the 1970s was foreshadowed by the fate of similar efforts at earlier moments of crisis. Resolution of these earlier struggles had left an institutional legacy that affected the later crisis. In particular, the scarcity and fragmentation of bureaucratic expertise and operational capabilities provided few bases on which to build new government powers and responsibilities. Precedents for direct state involvement in the financing and production of energy, and constituencies in support of these activities, had not been forged in earlier historical periods. In the absence of such precedents, state-building efforts were overwhelmed by opposition.

By the end of the decade, executive officials had not greatly expanded their institutional capabilities to shape the course of energy adjustment. In ways that are explicable in historical and comparative perspective, state building would not be a viable response to the energy upheavals of the period. For executive officials concerned with energy adjustment, policy necessarily had to turn to more readily available capabilities.

[81]See Edward Cowan, "Washington's Role in Oil Industry Stirs Controversy," *New York Times*, February 5, 1979, pp. 47, 57.

CHAPTER SIX

The Limits of Technology and Spending

> Science was an asset of state, not only because it furnished new tools
> and improved techniques of war, but because it contributed directly
> and indirectly to the general prosperity, and prosperity contributed
> to power.
>
> David S. Landes, 1969

As international adjustment schemes and state-building efforts
were breaking down, another policy effort was picking up mo-
mentum. It was a wide-ranging set of spending programs aimed at
stimulating energy production, dampening energy consumption, and
laying the groundwork for a national transition from traditional to
advanced and renewable energy sources. In effect, executive officials
fell back on a traditional form of public policy: spending programs
designed to alter or encourage private behavior. In this case the
spending instrument was tied to an energy science and technology
policy, creating an even more compelling rationale for the disbursement of public funds. For a program presented as a forward-looking
national investment in the transition from petroleum to advanced
forms of energy, the public purse gradually opened, and by the end
of the decade the American federal government had massive spending programs for massive research and development.

The attractiveness of these spending programs for administrative
officials otherwise hemmed in by government institutions is clear.
Expenditure programs required less centralized organization and did
not raise the threat of direct government involvement in private commercial affairs. Although they promised little in the way of immediate
relief, they did provide opportunities for modest levels of govern-

ment leadership. Unable to alter prevailing government roles in the energy area, administrative officials resorted to pushing more money through current institutional channels.

Use of the spending instrument took the form of public expenditures for research and development. Executive officials did not think R & D programs promised simple solutions to energy adjustment problems, but they did think they provided a cutting edge for state leadership on energy adjustment. Because the American state could get involved in encouraging the development of technology, the energy R & D program provided a basis for leadership.

There are three reasons why the state could create this leadership role for itself in the 1970s. First, there was a "policy space" in which government could take on a leading role. There was no great resistance, and indeed, scientists and various congressional officials were actively encouraging organizational and programmatic expansion of the federal government's efforts in science and technology.

Second, the fundamental logic of energy R & D initiatives involved spending money. Other federal initiatives were blocked, but the disbursement of money was a readily available option. Although the American government had little direct power to alter patterns of energy consumption and production, it could spend money—and in so doing indirectly encourage new patterns.

Third, the science and technology option involved a small circle of political players. Congressional committees were willing to appropriate huge sums of money, and planning did not require a large organizational infrastructure. Science and technology experts could be recruited into the executive branch to act as program managers, supervising the spending programs to encourage technological development. In essence, the government could exercise its new role without challenging established societal interests or exhorting society to adopt new forms of behavior.

ORGANIZATION OF ENERGY RESEARCH AND DEVELOPMENT

In the years prior to 1973 no government agency had central control over energy R & D. Various public and private studies and reports in the postwar period proclaimed the importance of a unified government role in energy science and technology, and in the early 1970s the urgency of these exhortations quickened. Nonetheless, federal responsibility for energy R & D remained bureaucratically fragmented and unprepared to generate a national strategy.

In the absence of a national R & D strategy and a central organization to undergird it, government expenditures remained the tool of disparate bureaus and agencies. One government study noted that without a national policy, "the method for managing those [R & D] activities requiring Federal attention has evolved gradually and in an uncoordinated fashion." Consequently, "energy R & D has been conducted by the different agencies as a part of their particular missions."[1]

Before 1968 modest coordination was possible only through the budgetary process. As one congressional study notes, "in the yearly review by the then Bureau of the Budget (now the Office of Management and Budget) of individual agency budget requests, some attempt was made to assure that the mission-oriented requests did not overlap requests for funds for the same programs from other agencies with other mission." However, it was difficult for the bureau to do more than prevent simple duplication. "There was no rational way for them to determine whether the levels requested were too high or low, whether there was balance among the programs, or whether important omissions were occurring."[2]

With the establishment in 1968 of the Energy Policy Office in the Office of Science and Technology, coordinating and planning efforts were attempted. Some achievements were made in specific projects; also, the office attempted to rationalize and balance the multiple bureaucratic bases for the sponsorship of energy R & D. However, the tasks of planning and coordinating continued to overwhelm the modest, three-person staff of the office. With the office directing its efforts at ad hoc problem solving, coordination had to remain with the budget mechanism.

Early Organizational Problems

The problem of planning and coordination should have come as no surprise for officials of the Nixon administration. The organizational and analytic inadequacies confronted in the early 1970s had been identified a decade before, in the first systematic study of energy R & D. An interdepartmental task force, initiated by the Kennedy administration and reporting in 1964, had identified problems in existing federal activities in energy R & D. The report acknowledged that energy R & D was carried on "at so many levels and in so many

[1]"Energy Research and Development—Problems and Prospects," Committee Print prepared for the U.S. Senate, Committee on Interior and Insular Affairs, 93d Cong., 1st sess., p. 47.
[2]Ibid., p. 48.

different types of institutions that it is not easy to derive a composite picture of what is being done, what is being neglected, and what should be started now if a possible future crisis is to be averted." Nonetheless, although abundant domestic resources were adequate for the short term, longer-range requirements made necessary an expanded and coordinated R & D program. Indeed, the central failing of the existing government effort, the report concluded, was the absence of a "long-range, integrated plan for civilian energy R & D." This institutional problem led to others, such as delays in initiating essential research with distant payoffs and inadequate support for programs without demonstrable high promise.[3]

The report identified a broader problem, one that would finally be redressed a decade later: the paucity of professional experts in energy research, both within and outside government. It concluded: "The scarcity of R & D talent in some disciplines, in fact, prevents full implementation of many essential programs, and it is critical for certain long-range research projects that must be initiated promptly if the payoffs are to ensue within the necessary time scale." A unified field of "energy research" has not existed. Rather, energy research has been carried out in disparate and unconnected scientific subfields. Each type of research demands a different institutional setting. For these reasons, the report argued, a "dynamically balanced program in civilian energy R & D" would require more than a laissez-faire approach. It "must include appreciation of the necessity for the Government to assume a new role—stimulus, catalyst, or sponsor, as appropriate—if implementation of the overall program is to be assured."[4]

This analysis foreshadowed the R & D imperatives rediscovered in the early 1970s. The report illustrates the professional view that institutional innovations were needed to facilitate long-term planning and investment. However, innovation would have to await the crises of the 1970s.

CONGRESS AND THE SCIENTIFIC COMMUNITY

The initiatives taken by the Nixon administration in the early 1970s were not simply matters of narrow concern to executive officials.

---

[3]U.S. Executive Office of the President, Office of Science and Technology, *Energy R & D and National Progress*, Prepared for the Interdepartmental Energy Study by the Energy Study Group under the direction of Ali Bulent Cambel (Washington, D.C.: GPO, June 5, 1964), pp. xviii–xix.
[4]Ibid., pp. xviii–xxi.

Proposals emerged in a political context highly disposed to support an enhanced federal role in R & D. This support, which created a favorable policy space for organizational reform, came both from Congress and from the scientific and professional research community.

Support came at two levels. Various groups and agencies were advocating higher levels of spending on energy and R & D and more coherent planning techniques. More generally, a large array of groups and officials came to urge a broad-scale upgrading of the federal government's commitment to science and research.

The first sort of professional research concern for an expansion of government responsibility for energy R & D can be seen in a 1971 report prepared for the National Science Foundation. The report was largely a technical discussion of state-of-the-art opportunites in energy R & D; however, criticism of existing administrative and funding commitments formed a central theme.[5] The report concluded that in the United States, energy-oriented R & D, both public and private, was at an intolerably low level.

The view that government was not adequately organized for the effective deployment and direction of science and technological resources was shared more broadly. The science and research community, that is, professional scientists and engineers within and outside government, also called for a strengthened government presence in science and technology affairs. This concern was expressed in the National Science Foundation. In this body, for example, new programs were proposed in invigorate government support for science and technology development. In 1973 the NSF proposed a National R & D Assessment Program and an Experimental R & D Incentives Program. These programs were to be funded in the following year at the level of $30 billion.

Much of this concern from the professional research community was triggered by Richard Nixon's abolition of the science policy machinery within the Executive Office; responsibilities for advice on civilian R & D programs had been transferred to the National Science Foundation. These actions prompted concern over an issue that had surfaced periodically throughout the postwar period: where should science and technology policy be located within the executive policy-making establishment, and how important was it?

In 1973 the National Academy of Sciences reaffirmed the need for a high-level apparatus to advise the president.[6] It proposed a council

[5]See National Science Foundation, Summary Report, *The U.S. Energy Problem,* prepared by the Inter Technology Corporation (Washington, D.C., November 1971).
[6]"Science and Technology Policymaking—A Proposal" (Washington, D.C.: National Academy of Sciences, 1973).

of advisers similar to the abolished Office of Science and Technology and which was to function not unlike the Council of Economic Advisers.

Other calls for reform were registered before Congress. One important voice was that of S. David Freeman, formerly an official in the Nixon administration's Office of Science and Technology. In congressional hearings in 1972 Freeman spoke as director of the Ford Foundation's Energy Policy Project. He expressed a view shared widely among government, industry, and other private groups, that the nation needed a stronger federal effort in energy R & D. He argued that major areas of energy R & D remained grossly neglected and that this was a central reason for the serious national energy crisis then emerging. Behind inadequate government efforts was an organizational problem:

> I am, of course, very much aware of the numerous Federal agencies and private companies engaged in energy research and development. With the exception of the Atomic Energy Commission the efforts are scattered and relatively weak. They are carried on in small offices in large departments. The department heads have numerous responsibilities which make it difficult for them to focus on energy R & D as a high priority item. There is no agency with responsibility to develop new sources of energy, such as solar, all the way to commercial feasibility. There is no agency to develop the means for consuming energy more efficiently, which I think holds great promise in helping us match supply and demand without shortages. There is no agency to compare all the rich array of options or push for a strong across-the-board program with commitments to produce commercial hardware on a realistic timetable.[7]

Later, Freeman pointed to the need not just for organizational reform but also for the recruitment of experts into central administration. He argued that "in order to lead a new thrust in energy R & D the responsible Federal agency must possess a critical mass of technical talent. Without such talent the program would be at the mercy of promoters and special interests." Although the executive branch already had many experts, "the necessary depth of technical talent to carry major projects to commercial success does not seem to me to exist anywhere in the Federal energy establishment outside of the AEC."

Other members of the national science establishment advocated integrated energy R & D programs and an enhanced federal organi-

[7]"Energy Research and Development," Hearings before the Subcommittee on Science, Research and Development of the Committee on Science and Astronautics, U.S. House, 92d Cong., 2d sess., May 1972, p. 299.

zation. One prominent member of the science community was MIT scientist James R. Killian, Jr., who in 1972 congressional testimony had also signaled the need for federal action. "I support the view that the construction of a sound energy program for the Nation calls out for new organizational arrangements with the Federal Government, and for new ways to select the policies that will concert our actions and clarify our goals," Killian argued. "I also want to emphasize that such a well conceived program will require policy guidelines and further integration of institutions for executing policy if we are going to do it well."[8]

Representatives from the science community clearly wanted to remain at the center of federal initiatives. In 1972 a National Science Board report "The Role of Engineers and Scientists in a National Policy for Technology," recommended that scientists and engineers gather in special groups "to explore specific large problems of national importance, and to explore alternatives for dealing with those problems." It also suggested that these specialists could "put forward a menu of alternative solutions from among which choices can be made by the established decision making process of Government." The NSB did not envisage these exploratory groups as replacing political officials in the political process. Rather, the groups were to frame choices and suggest possibilities.[9]

This report summarized the views of the science community on federal involvement in energy R & D and sought a greater role for government. To facilitate this expanded role, the community recognized, new institutions would be needed to coordinate and rationalize funding and administration. The NSB and those it represented hoped that those new institutions would build in major participation by the scientists and engineers themselves.

Congressional Initiatives

Other support for new federal efforts in science and research came inside Congress itself. Congressional interest in energy R & D came from many quarters, but it was concentrated in a subcommittee of the House Committee on Science and Astronautics. In 1970, after several months of hearings, the Subcommittee on Science, Research, and Development released a report recommending an expanded and integrated science policy. The subcommittee underscored the impor-

[8]Ibid., pp. 300, 5.
[9]National Science Foundation, National Science Board, "The Role of Engineers and Scientists in a National Policy for Technology" (Washington, D.C.: GPO, 1972).

tance of science and technology to social problem solving. It urged a national program with formal policy goals and with planning responsibilities concentrated in a single administrative agency.[10]

This subcommittee, in the following year, focused more specifically on energy R & D. Led by Representative Mike McCormick, the subcommittee formed a task force on energy R & D, and it began meetings and briefings in 1971. In May 1972 the subcommittee held hearings, and the task force issued a report in January 1973. Along the way the task force published an inventory of energy research. In the 1973 report the task force urged that a "greatly increased national energy research and development effort" be implemented. Furthermore, it recommended organizational reform. The White House must be the "focal point" for the new effort, and a single operating agency should be established to support energy R & D. Finally, the report suggested particular priorities for federal funding.[11]

The full House Committee on Science and Astronautics continued to conduct hearings and issue reports during the 1970s. The committee held hearings in 1973 and 1974 whose theme was that new social and economic problems were producing new imperatives for high-level policy making in matters of science and technology.[12] In 1973 the newly formed Subcommittee on Energy conducted hearings focused directly on substantive aspects of energy R & D. Representative McCormick, as subcommittee chairman, stressed "the high priority which I feel should be assigned to energy R & D." In testimony from administration officials, the subcommittee sought to extract commitments from the administration for budget increases and specific targets for energy research.[13]

The second major congressional source of support for expanded programs for energy R & D was to be found in the Senate. In May 1971 the Senate approved a long-term study of energy and resource problems, to be conducted under the sponsorship of the Senate Com-

[10]"Toward a Science Policy for the United States," Report of the Subcommittee on Science, Research, and Development to the Committee on Science and Astronautics, U.S. House, 91st Cong., 2d sess., October 15, 1970.

[11]"Energy Research and Development," Report of the Task Force on Energy of the Subcommittee on Science, Research, and Development of the Committee on Science and Astronautics, U.S. House, 92d Cong., 2d sess., December 1972, pp. 3–5.

[12]"Federal Policy, Plans, and Organization for Science and Technology," Hearings before the Committee on Science and Astronautics, U.S. House, 93d Cong., 1st and 2d sess., Part I, July 1973; Part II, June–July 1974.

[13]See, e.g., "Energy Research and Development—An Overview of Our National Effort," Hearings before the Subcommittee on Energy of the Committee on Science and Astronautics, U.S. House, 93d Cong., 1st sess., May 15, 1973. In this session National Science Foundation officials were questioned concerning funding and commitments.

mittee on Interior and Insular Affairs.[14] The mandate was quite broad, to study resource availability and emergent energy problems, to explore the impact of federal laws and government programs on the energy industry, and to recommend legislation. Senator Henry Jackson, chairman of the committee, was charged with responsibility, and in the years that followed this committee responsibility placed Jackson at the center of energy policy debate. The study produced a myriad of reports and hearings. These materials continued to appear throughout the Nixon and Ford administrations, although they never became a well-tailored legislative agenda or energy program.[15]

R & D AND THE NIXON PROPOSALS

The first national call by a president for a systematic policy for energy R & D came in June 1971. In a message to Congress, President Nixon outlined a general program to meet what was perceived to be a gradually emerging energy challenge. Nixon committed government to "providing technical leadership" on a broad range of emerging technologies and sources of energy. The president claimed "that the time has now come for government and industry to commit themselves to a joint effort to achieve commercial scale demonstrations in the most crucial and most promising clean energy development areas—the fast breeder reactor, sulfur oxide control technology and coal gasification."[16] R & D initiatives encouraged by the government were to be a central response to the dislocations and shortages caused by rapidly expanding energy consumption.

This new commitment to research and development was to be facilitated by organizational reform. Acknowledging the problems of bureaucratic dispersion of responsibility for research and policy, Nixon proposed that programs for energy production and development be consolidated into a new Department of Natural Resources. He argued that "one of the major advantages of consolidating energy responsibilities would be the broader scope and balance this would give to research and development work in the energy field."[17]

[14]"Authorization for a Study of National Fuels and Energy Policy," *Congressional Record*, May 3, 1971, S13227–13230.

[15]See Neil De Marchi, "The Ford Administration: Energy as a Political Good," in Craufurd D. Goodwin, ed., *Energy Policy in Perspective* (Washington, D.C.: Brookings, 1981), pp. 489–90.

[16]"Message to the Congress on a Program to Insure an Adequate Supply of Clean Energy in the Future," June 4, 1971, rpt. in "Presidential Energy Statements," Committee on Interior and Insular Affairs, U.S. Senate (Washington, D.C., 1973), p. 3.

[17]Ibid., p. 11.

The Nixon administration announced it would seek to consolidate bureaucratic functions into centralized superdepartments in the 1971 State of the Union Message. A Department of Natural Resources, it was argued, would concentrate fragmented government responsibilities. Within this new Department an Energy and Mineral Resources Administration would centralize and direct energy R & D programs, creating the organizational bases for a strategic R & D policy. "Broader scope and greater balance would be given to nationally supported research and development work in the energy field."[18]

In March 1972 the president again addressed himself to the need for new government capacities. Nixon specified a variety of conditions—the sheer size of projects, their expense, and their risks—that justified the presence of government leadership. Secretary of Commerce Peter G. Peterson, in congressional testimony later that month, elaborated on the view of the executive branch. After noting the president's stress on research and technology as instruments of public problem solving, Peterson focused on energy: "We have identified an energy crisis in this country, and we have realized that one crucial way to resolve this crisis is through technological advances."[19]

1973 Nixon Proposals

The symbolic overture for the R & D initiatives came in June 1973, with a new set of proposals from the Nixon administration. The centerpiece was a proposal for a $10 billion program for energy R & D spread over a five-year period beginning in Fiscal Year 1975. In addition to this projected spending, Nixon authorized $100 million in the current budget to be "devoted to the acceleration of certain existing projects and the initiative of new projects in a number of critical research and development areas." The presidential message directed the chairman of the Atomic Energy Commission (AEC) to review prevailing efforts by government and industry in energy R & D and to recommend an integrated national program. The president noted

---

[18]State of the Union message, January 22, 1971, in *Weekly Compilation of Presidential Documents*, vol. 7, January 25, 1971, pp. 94–95. The proposals were transmitted to Congress on March 25, 1971. Quotation from *Papers Relating to the President's Departmental Reorganization Program: A Reference Compilation* (Washington, D.C., March 1971), p. 170.

[19]"Science, Technology, and the Economy," Hearings before the Subcommittee on Science, Research, and Development of the Committee on Science and Astronautics, U.S. House, 92d Cong., 2d sess., April 11, 1972, p. 12.

that "this program should encourage and demonstrate new technologies that will permit better use of our energy resources."[20]

The June message also contained new organizational proposals. Nixon proposed a new, independent agency, the Energy Research and Development Agency (ERDA), to "focus" all federal energy R & D by merging the R & D responsibilities of the Atomic Energy Commission and the Department of Interior. The agency was to have responsibility for the planning and management of government operations and expenditures. In addition, he again proposed a Department of Energy and Natural Resources to take over authority for energy policy from the Department of Interior and other agencies. Finally, the president acted immediately to upgrade the existing energy policy staff in the Office of the President, by expanding existing staff into the Energy Policy Office. An advisory council of outside experts would advise this office.[21]

The initiatives led to little direct action, for the administration was still engaged in a complex reorganization within the executive branch. As late as April 1973, for example, in a message to Congress on energy, the president was still seeking to strengthen energy R & D within the Department of Interior. In addition, there were problems in the continuing absence of analytic capacities to evaluate the merits of particular programs and a centralized agency capable of setting research priorities.[22]

The Nixon administration, in the two years preceding the October embargo, had been developing broad policy proposals to meet energy problems that were not yet understood as an immediate crisis. On the one hand, these efforts were part of a larger executive plan for bureaucratic reform—embodied in the proposal for a Department of Energy and Natural Resources.[23] This proposal met with stiff congressional resistance, but other initiatives, more directly related to energy R & D, were unfolding. The most important organizational step in this regard came with the June 1973 proposal for a centralized

[20]"Statement Announcing Additional Energy Policy Measures, June 29, 1973," rpt. in "Presidential Energy Statements," Committee on Interior and Insular Affairs, U.S. Senate, 1973, pp. 52–53.

[21]Ibid., pp. 57–58.

[22]"Message to the Congress, April 18, 1973," in ibid., p. 26. During 1972 and early 1973 analysis and coordination of energy R&D policy came from the Office of Science and Technology (before it was abolished), assisted by the Federal Council for Science and Technology. "Energy Research and Development—Problems and Prospects," Committee on Interior and Insular Affairs, U.S. Senate, 1973, p. 47.

[23]Neil de Marchi, "Energy Policy under Nixon: Mainly Putting Out Fires," in Goodwin, *Energy Policy in Perspective*, p. 412.

Table 5. Government expenditures on non-nuclear energy research and development, in 1981 U.S. $ millions

	1976	1977	1978	1979	1980
West Germany	116.7	152.9	239.1	282.9	261.1
United States	890.1	1591.9	1927.8	2267.7	2620.4
Japan	270.9	109.3	118.1	137.6	361.6

Source. Calculated from data in International Energy Agency, *Energy Research, Development and Demonstration in the IEA Countries* (Paris, 1982).

energy R & D agency, the Energy Research and Development Administration. This agency would centralize energy R & D responsibilities and generate a national strategy for implementation.

The Energy Research and Development Administration

By 1975 the organizational machinery was in place and the disbursement of R & D money had begun. As Table 5 and Figure 2 indicate, government expenditures rose dramatically after 1974. Between 1974 and 1977, for example, energy-related R & D rose 191 percent, and the increases are even more striking when nuclear (fission and fusion) research funds are factored out. In a discussion of the energy R & D expenditures of different governments, the National Science Board noted in 1981 that "the United States is . . . one of the few OECD countries which have significantly and consistently increased public financing of energy research and development since the energy crisis began in 1973. The U.S. energy R & D budget rose 469 percent from $0.6 billion in 1973, to $3.6 billion in 1980."[24]

Throughout the 1970s energy expenditures were one of the fastest-growing parts of the federal budget. Several Department of Energy research analysts have indicated that few budget requests were left unfunded, and only rarely were the substantive aspects of particular projects challenged. "I had essentially a blank check," one official summarized.[25] The political environment for the growth of government support for energy R & D was highly favorable. Few groups or officials opposed generous funding of research projects.

[24]National Science Board, *Science Indicators 1980* (Washington, D.C.: GPO, 1981), p. 10.
[25]Interview, Department of Energy research official, March 14, 1983.

Figure 2. Federal funding for energy research and development programs, 1971–83

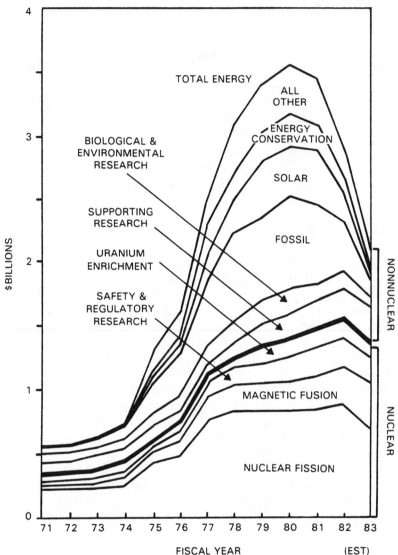

SOURCE. National Science Board, *Science Indicators, 1982* (Washington, D.C.: National Science Foundation, 1983), p. 244.

Organizational Development

The bureaucratic handling of these huge budget requests came to reside in ERDA, which was established in 1975. ERDA received administrative and congressional support to build a large-scale organization for planning and operations. Congress, in funding the first ERDA budget, stressed the technological dimension of the "Nation's energy challenge," claiming that energy solutions "will require commitments similar to those undertaken in the Manhattan and Apollo projects."[26]

The agency recruited scientists and economists and became the main government body generating proposals for targeted research. In its first years ERDA was much envied; as one observer at the time noted, ERDA had "plenty of money and a lot of glamour," and scientists and various experts from around the executive establishment were eager to move to ERDA. Chief administrator Robert Seamans also recruited officials from private industry. Behind the ambitiousness of the ERDA mission and the esprit of its early incumbents was Congress, which lavished funds on the new organization. "Congress believes that R & D, at least, is something the American government knows how to do," one observer noted.[27] With this political support and the ongoing problems of energy adjustment, ERDA became a highly visible and active bureaucratic agency within the executive establishment. By mid-1975 it had put forth a comprehensive plan for energy R & D, a plan that provided rationales and priorities to guide the large budget expenditures that Congress was beginning to pass.[28]

ERDA brought together the major federal responsibilities for energy R & D. Personnel and resources were transferred from the Atomic Energy Commission, bringing in nuclear research, the national laboratories, and military nuclear weapons research. Facilities from the Interior Department's Office of Coal Research and the Bureau of Mines, and a variety of National Science Foundation programs concerned with solar and geothermal research, were incorporated into ERDA. Finally, programs in the Environmental Protection Agency concerned with advanced automotive research were brought under the new research umbrella. A total of 7,222 employees were trans-

[26]"Federal Nonnuclear Energy Research and Development Act of 1974," Public Law 93-577, 93d Cong., December 31, 1974, sec. 2(c).
[27]Tom Alexander, "ERDA's Job Is to Throw Money at the Energy Crisis," *Fortune*, July 1976, p. 153.
[28]ERDA 48, *A National Plan for Energy Research, Development and Demonstration*, 2 vols., June 28, 1975. This original plan was revised and submitted to Congress in early 1976 as ERDA 76-1, *A National Plan for Energy Research, Development and Demonstration: Creating Energy Choices for the Future*, 2 vols., April 15, 1976.

fered to ERDA, which began with an operating budget of approximately $3.6 billion.[29]

Planning Strategy

After its passage by Congress, ERDA was charged with developing a national plan for energy technology. With this general mandate ERDA produced a set of broad energy goals and developed a strategic planning approach to bring federal R & D resources to bear on them. It identified both conservation and production projects. Short-term, high-priority conservation investments included technologies to increase industrial and transportation efficiency. Among energy production goals, the important short-term priorities included technologies to facilitate the use of coal and enhance the recovery of oil and gas, as well as a continued commitment to light-water reactors.[30]

Longer-term commitments were made to energy technologies that expanded renewable sources, among them the breeder reactor, solar electric power, and wind and thermal power options. The immediate goal of ERDA was to move practical technologies to the commercialization stage quickly and then relinquish control to private industry. More remote technologies were to be encouraged at the level of basic research.[31]

Legislative mandates ensured that ERDA would focus its efforts on different energy sources. The Solar Heating and Cooling Act of 1974, the Geothermal Energy Research, Development, and Demonstration Act of 1974, and the Solar Energy Research, Development, and Demonstration Act of 1974 instructed ERDA to develop programs for effective use of particular types of energy.[32]

The most important aspect of ERDA's mission was the establishment of program planning and implementation. Initially ERDA committed itself to the development of analytic techniques for program building. Planning responsibilities for R & D were carried out within the framework of a Planning, Programming, Budgeting, and Review (PPBR) system. This system, as ERDA described it, was "to provide an

[29]Department of Energy, Historical Division, "History of the Energy Research and Development Administration" (Washington, D.C., March 1982).

[30]On the general plan see Federal Nonnuclear Energy R & D Act of 1974, Public Law 93-577; the administrative authority for this planning responsibility came from the Energy Reorganization Act of 1974, Public Law 93-438. The law mandated that the plan be updated annually. On short-term priorities see ERDA 76-1, 1:25.

[31]ERDA 76-1, 1:35–68.

[32]Department of Energy, Historical Division, "A History of the Energy Research and Development Administration," March 1982, p. 3.

integrated and disciplined approach to analyzing the Nation's future energy technology needs; formulating the Federal role in addressing those needs; designating targeted programs to conduct ERDA's portion of the Plan; allocating resources consistent with the Plan and program design; and ensuring that ERDA's programs are effectively managed."[33]

At the center of the ERDA planning system was an assessment process to pinpoint projects for funding. The "strategic planning logic" is detailed in Figure 3. Its fundamental intent was to provide systematic assessments of opportunities for government intervention in energy R & D. Government involvement would be justified where returns on private investments were inadequate yet public returns on investment were deemed sufficiently high. Where public goals were involved, the planning process was to specify appropriate government mechanisms for implementation and review. Finally, the planning process was to be built upon continually enhanced analytic capacities. "An integral part of this Plan is a detailed program for improving the informational base for these assessments, facilitating ERDA's access to this information, and developing the tools to better analyze the implications of new technologies in terms of economic growth, environmental impact, and public policy."[34] One analyst described the strategic planning more straightforwardly: "It will determine which of [ERDA's] goals already meet private-sector investment criteria. For those that don't, ERDA will design whatever incentives (e.g., guaranteed loans, capital grants, price supports, research and development funding) are required to attract private investment."[35]

ERDA was molded into an active bureaucratic organization by its first administrator, Robert C. Seamans, Jr. Formerly president of the National Academy of Engineering, Seamans brought ERDA to the leading edge of energy planning during the mid-1970s. His belief was that government had a special responsibility for moving the nation through a historic energy transition and that this contribution would center on science and technology. In his confirmation hearing Seamans underscored his "strong conviction that science and technology can and must make vital and fundamental contributions to both the short and long term solutions to our national energy predicament." To this end Seaman affirmed his belief in a "strong R & D agency

[33]ERDA 76-1, 1:81.
[34]ERDA 76-1, pp. 82–84; quotation at p. 13. See also "A Review of the Energy Research and Development Administration's National Energy Plan," prepared for the Subcommittee on Energy Research and Water Resources, Committee on Interior and Insular Affairs, U.S. Senate, 94th Cong., 2d sess., December 1976.
[35]Alexander, "ERDA's Job," p. 162.

Figure 3. Strategic logic for ERDA planning decisions

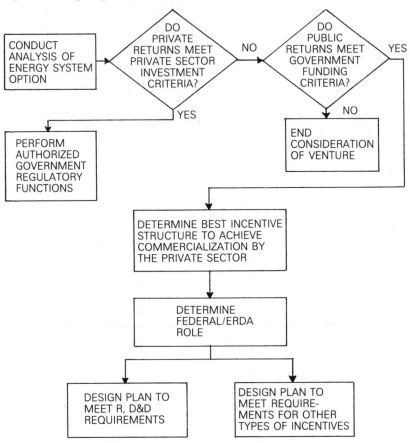

SOURCE. ERDA 76–1, "A National Plan for Energy Research, Development & Demonstration: Creating Energy Choices for the Future, 1976" (Washington, D.C., 1976), 1:83.

capable of developing and sustaining a balanced and practical program for energy generation and conservation."[36] With the ERDA

[36]"Nomination of Dr. Robert C. Seamans, Jr. to Be Administrator, Energy Research and Development Administration," Joint Hearings before the Joint Committee on Atomic Energy and the Committee on Interior and Insular Affairs, U.S. Senate, 93d Cong., 2d sess., December 11, 1974, pp. 8, 7. The Department of Energy's Historical Office notes: "Seamans believed that a sixty-year lead time was no longer possible as in past energy transactions, and that in the current situation a transition to new forms of energy would have to be made in half the time and in a far more complex world." "A History of the Energy Research and Development Administration," p. 4.

plan as his organizing document, Seamans became active in the policy process.

Within the executive branch Seamans met weekly with the Ford administration's Energy Resources Council (also authorized by the Energy Reorganization Act of 1974) to coordinate ERDA plans with the government's larger efforts on energy policy. After the first ERDA plan took shape, Seamans and other officials conducted regional public discussions.[37] The most important political relationships that Seamans and ERDA forged, however, were with relevant congressional committees. Here ERDA was pressed to relate its efforts to the formal administrative goals of energy self-sufficiency. In particular, although Congress tended to support most R & D funding requests, it did demand better indicators for program evaluation. Senator Jackson, for example, who chaired the Committee on Interior and Insular Affairs, asked for more sophisticated program analysis, pointing out that "it is the Committee's feeling that there is a serious need for more specificity in the plan [ERDA 76-1] through targets and costs for the various plan elements. Any plan requires some quantitative measure of success, that is, clearly defined targets that allow others to judge the effectiveness of the actions taken to achieve the planned goals." Seamans responded that ERDA was "still developing the techniques and information necessary to improve our procedures for allocating resources among the many energy R & D options." But he also indicated the limits of strict cost-benefit analysis caused by "uncertainties derive[d] primarily from the forecast sensitivity to factors such as energy policy, government programs, economics of technology, and market situations."[38]

Despite the analytic limits on evaluation of its programs, ERDA generated three national R & D plans between 1975 and 1977. The Department of Energy's Historical Office summarizes these efforts as follows: "ERDA . . . made significant progress in developing national energy research and development plans, in mobilizing talent, and in coordinating the diverse energy activities formerly scattered among many federal agencies. The staff worked closely with all segments of industry, with the academic community, with foundations, with nonprofit corporations, and with foreign countries. . . . [B]y the time

[37]See DOE, "History of the Energy Research and Development Administration," pp. 4–5.
[38]Letter reprinted in "A Review of the Energy Research and Development Administration's National Energy Plan," prepared for the Subcommittee on Energy Research and Water Resources, Committee on Interior and Insular Affairs, U.S. Senate, 94th Cong., 2d sess., December 1976, pp. 27, 32.

ERDA was absorbed into the Department of Energy in the fall of 1977 programs were well underway for the near-term demonstration of more efficient ways to recover and use coal, oil and shale, and a number of pilot plants had been constructed or were in progress."[39]

ERDA was an attempt to construct an administrative apparatus capable of identifying and funding emerging energy technologies. It is revealing that Seamans, its first administrator, came from a background in large-scale technology projects, most notably the Apollo Project. Seamans drew strong parallels between ERDA and NASA. In 1977 Seamans noted that these two major projects "are the largest nonmilitary, government-run research and development organizations ever established." He went on to note that "partly because of perceived threats from abroad, both commenced operations within an environment of strong presidential, congressional and public support."[40] The organization at its beginning was strongly imbued with an ambitious and vital administrative mission.

To its administrators, ERDA was to provide a focus for bringing together diverse and complex technology programs. Those officials, while attempting to design a coherent and autonomous government presence in the energy sector, recognized the inherent limits on ERDA's control. Administrative goals were generally long-term, and the fruits of ERDA programs were ultimately to be given to the private sector. These factors, and the complex relationships that had to be forged with universities and private industry, pulled at the centralization and strength of ERDA's leadership.

OMB and the 1977 Budget

Above ERDA (and later the Department of Energy) was the Office of Management and Budget, the instrument that related proposals being generated within ERDA to presidential goals. The OMB in many respects conforms to the archetypal "state organization"— isolated from interest groups and staffed by experts charged with making neutral judgements about budget proposals. Thus it can become a formidable counterweight to the play of particularistic interests. David Dickerson and David Noble note a relationship between the Office of Science and Technology and OMB that resembles the relationship between ERDA and OMB—an institutional pattern, they

[39]DOE, "History of the Energy Research and Development Administration," p. 14.
[40]Robert C. Seamans, Jr., and Fredrick I. Ordway, "The Apollo Tradition: An Object Lesson for the Management of Large-scale Technological Endeavors," *Interdisciplinary Science Reviews* 2 (December 1977), 271.

argue, that created "legitimation for the modern, executive-centered, interventionist state."[41]

Yet in addition to this role as counterweight, OMB has also been used to rank and aggregate proposals, allowing the president, where he perceived a special issue (as Ford and Carter did with energy), to impose his priorities on the budget process. Indeed, one study of science and technology policy concludes that during the 1970s the "OMB became a potent manager of science through the mechanism of the annual federal budget cycles."[42]

The ascendancy of the OMB as organizer and priority setter for energy research and technology initiatives was explicitly reflected for the first time in January 1977, in an important budget analysis document. In its major document for the 1978 budget, OMB provided the intellectual rationale for the federal government to take the lead in the nation's achieving energy adjustment goals, and for the special role of R & D funding as an instrument by which the executive branch might impose order and direction on that process.[43] In an otherwise conflict-ridden and decentralized policy process, funding for science and technology programs for energy was to provide the mechanism for shaping national priorities.

CARTER AND THE DISPERSION OF FUNDS

With the Carter administration energy R & D funding reached its zenith, both in level of expenditures and in diversity of targets. Indeed, the pattern of expenditures evolved over the Nixon, Ford, and Carter administrations. Early post-1973 funding, proposed in the context of the ambitious goals of Project Independence, focused on "big-ticket items" that would further efforts to expand domestic energy production rapidly. In subsequent years expenditures diversified, based increasingly on the assumption that domestic petroleum prices

[41]See Hugh Heclo, "OMB and the Presidency: The Problem of Neutral Competence," *Public Interest* 38 (Winter 1975), 80–98, and Larry Berman, *The Office of Management and Budget and the Presidency, 1921–1979* (Princeton: Princeton University Press, 1979). Quotation from David Dickerson and David Noble, "By Force of Reason: The Politics of Science and Technology Policy," in Tom Ferguson and Joel Rogers, eds., *The Hidden Election* (New York: Pantheon, 1981), p. 267.

[42]Deborah Shapley and Rustom Roy, *Lost at the Frontier: U.S. Science and Technology Policy Adrift* (Washington, D.C.: Brookings, 1985), p. 55.

[43]Office of Management and Budget, *Issues '78: Perspectives on Fiscal Year 1978 Budget* (Washington, D.C.: GPO, 1977), pp. 81–110. See also Claude E. Barfield, *Science Policy from Ford to Reagan: Change and Continuity* (Washington, D.C.: American Enterprise Institute, 1983).

would have to rise to the international level. It was the gradual recognition among executive officials that payoffs from the massive R & D effort were tied to energy decontrol and domestic price increases which added political force to the effort to deregulate.

Gerald Ford expanded energy research and technology budgets, considerably broadening coal and conservation programs. Jimmy Carter also diversified research and was more willing to fund such renewable sources as solar and geothermal. Types of programs shifted, from big and ambitious projects to longer-term and more marginal programs, precisely the types of programs that depended on a progressive imposition of market adjustment.

Under Carter, energy R & D expenditures gradually began to deemphasize nuclear research and strengthen research programs in non-nuclear areas. The final Ford budget proposal (1978) would have provided 64 percent of the total energy R & D budget to nuclear research; Carter changed it to 57 percent. By the end of Carter's administration, nuclear research amounted to only 41 percent.[44]

The state's role in R & D began to move closer to commercial demonstration. The Carter administration repeated the familiar guidelines that defined the proper role of government in energy R & D—government was not to drive out private investors or displace the resources of industry. Nonetheless, Carter did allow federal programs to tangle with market demonstration and current industry investments. Carter states: "A major challenge is to demonstrate technologies that will enable us to substitute [coal and oil shale energy sources] for our ever increasing oil imports. My program provides for the government to work closely with American industry to accelerate the demonstration of commercial-scale technologies that show promise of entering feasibility and the economics of conversion processes."[45]

With the creation of the Department of Energy, the activities of ERDA were brought under the administrative control of a large cabinet department. The new department's hierarchy continued to press for a unified planning apparatus, and energy technology policy remained a key vehicle for this planning effort. Secretary of Energy James Schlesinger told the House Committee on Science and Technology in January 1978 that "we are moving to bring the concept of unified Executive Branch planning to a reality by pursuing the comprehensive approach that is needed. . . [This] approach will use the

[44]Figures from Barfield, *Science Policy*, p. 20.
[45]Quoted in Barfield, *Science Policy*, p. 21.

wide variety of tools at the Department's disposal to stimulate greater and better use of new technologies." The Carter administration, like its predecessors, stressed the need for the executive branch to gain control over emerging energy technology. "In the past," Schlesinger argued, "we believe R & D has been conducted more or less in terms of the imperatives of the developers, rather than the imperatives of the economy at large."[46] For the Carter as for the Ford Administration, technology policy became an attractive vehicle upon which to build energy planning and budgeting.

The Carter administration's energy technology policy was publicly presented in an impressive, sophisticated language of "time phasing" and investment payoff calculations. The chief outcome of the policy, however, was a broadening of the definition of programs and technologies that would receive federal funds. Indeed, Schlesinger admitted in congressional testimony in 1979 that the disbursement of energy R & D funds had not followed very closely from an analytic design. "There has been a habit," he said, "particularly after the 1973–74 embargo, to scatter money around on various technologies of interest." This dispersion of funds carried a policy logic; Schlesinger argued that "there is no single solution to our energy problems. . . As we adjust to these future problems, what we will have is a whole set of small solutions which will contribute to the totality of the United States' grappling with its energy problem."[47] By 1978, for example, the share of government R & D funds going to nuclear programs had fallen to a new postwar low. Even in the mid-1970s, energy programs and technologies other than nuclear power were claiming a healthy share of federal energy R & D budget (see Table 6).

The juxtaposition of an energy R & D strategy with congressional willingness to appropriate funds makes it difficult to determine where strategy leaves off and constituency politics begins. At the very least the two were not always incompatible. The Department of Energy strategy of encouraging a wide range of new technologies sat well with a Congress eager to please those who would receive funds. Beyond this, the administration—beginning with ERDA and thereafter with the Department of Energy—sought persistently to put together a grand strategy. As late as 1979 Carter's energy officials were still

[46]"1979 Department of Energy Authorization," Hearings before the Committee on Science and Technology, U.S. House, 95th Cong., 2d sess., January 25, 1978, p. 10, and January 24, 1978, p. 35.

[47]"1980 Department of Energy Authorization," Hearings before the Committee on Science and Technology, U.S. House, 96th Cong., 1st sess., February 8, 1979, p. 43, and "1979 Department of Energy Authorization," January 24, 1978, p. 29.

Table 6. Nuclear R & D as a percentage of government funding
for energy R & D, by countries

Countries	1975	1976	1977	1978
France	91	87	86	82
West Germany	89	88	84	74
United Kingdom	94	92	88	84
Japan	95	82	74	66
United States	44	44	43	40

SOURCE. National Science Board, *Science Indicators 1980* (Washington, D.C.: GPO, 1981), p. 10.

grappling with a programmatic apparatus that would generate a coherent R & D strategy. Schlesinger noted in that year that "the process of project definition before one starts the large expenditures of funds has hardly been done at all."[48] He added that the Department of Energy was struggling to develop the analytic competence to decide on funding levels and time horizons for various energy technologies. As ERDA officials had found in the mid-1970s, the disjunction between program strategy and expenditures was large. It was not atypical for Congress to appropriate more funds than ERDA or the Department of Energy had requested. The House Committee on Science and Technology, for example, was a center of government support for a massive effort in non-nuclear R & D. In 1976 that committee raised ERDA's budget request in the non-nuclear area by over a quarter of a billion dollars.[49]

Nonetheless, officials throughout the Ford and Carter periods labored to shift spending in one direction or another, acting according to an elaborate plan but implementing congressionally approved energy R & D programs that were rough around the edges. Charges of pork barrel politics were rife and not unfounded.[50] Yet the aim of the state was indeed to channel money into the private sector, with the hope of providing a more or less efficient stimulus for the development of energy. The American government had sought similar responses to socioeconomic crisis before, spending money as a sort of public bribery in the service of national goals. Although energy R & D officials sought to spend money according to strategic policy, govern-

[48]"1980 Department of Energy Authorization," February 8, 1979, p. 43.
[49]*Congressional Quarterly*, May 15, 1976, p. 1189; on the general point of political and analytic processes in science and technology policy see W. Henry Lambright and Albert H. Teich, "Policy Innovation in Federal R&D: The Case of Energy," *Public Administration Review* 39 (March/April 1979), 140–47.
[50]Alexander, "ERDA's Job."

ment could not implement policy "cleanly"—it was a cost of the type of policy instrument being used.

The rich irony of these technology spending programs was in the incentives they produced for a return to market pricing. Carter's officials began their tenure with an elaborate program to stimulate energy production and conservation without moving directly to decontrol oil and natural gas prices. The Crude Oil Equalization Tax epitomized this elaborate effort to solve energy problems without confronting a difficult and politically costly problem. However, the successful commercialization of investments in energy technology, built up over the course of the decade, eventually turned on the question of pricing. If traditional energy sources—chiefly oil and natural gas—continued to be subsidized through price controls, the new and competing energy technologies would remain at a disadvantage. By the end of the 1970s the accumulated size and diversity of these new technology programs weighed on the side of market pricing.

It was for this reason that in 1978 the secretary of energy began to talk more forcefully about market pricing. "The first step to . . . commercialization is to provide financial inducements to industry to move in that direction [toward new technologies]. That is the reason we are attempting to get the price of oil up to the world price level and to provide a pattern of incentives in rebates in order to induce the shift away from oil and natural gas toward coal." He concluded: "As long as we subsidize the price of oil in this country it is not going to be attractive to industry to shift."[51] In the later months of the Carter administration, therefore, while the funding of programs remained the central tool for the exercise of government power, all issues began to turn on the question of energy pricing and the market.

CONCLUSION

The spending instrument was used in a wide variety of ways after 1973. In this mixture of expenditure and taxing programs, energy R & D funding became a particularly important government instrument. Of course, the political usefulness of science and technology spending programs predated the energy problems of the 1970s. Indeed, programs of this sort reflected a more general conviction among executive officials and professionals that systematic science

[51]"1979 Department of Energy Authorization," January 25, 1978, p. 38.

and technology spending would give the government new opportunities for problem solving in critical areas of public policy. The
growth of R & D funding as a state *tool* has been noted by James Katz:

> During the 1950s, research was perceived as important, but largely as
> insurance for national defense and to a lesser extent as a means of
> fostering public health. The Johnson and Nixon Administrations gave
> increased attention to the uses of research for social needs. First, the
> social problems, then the economic problems of the nation were per-
> ceived as soluble by additional research. These ideas have slowly gained
> intellectual currency (although hard-currency allocations were often
> postponed due to more immediate budget constraints) over the last
> twelve years. The Ford administration was the first to enunciate this
> belief formally and it also backed this commitment with significant
> money.[52]

The most important expression of this increasingly systematic and
ambitious science and technology policy was in the area of energy
adjustment.

The problem-solving potential of science and technology policy was
affirmed by the Nixon administration's secretary of commerce, Peter
G. Peterson, in congressional hearings in 1972. He cited energy problems as a priority area. Peterson argued that "government can play a
crucial role in the research and development of new ideas for solving
our growing social problems. . . We have identified an energy crisis in
this country, and we have realized that one crucial way to resolve this
crisis is through technological advances. That is why the fiscal year
1973 budget had a major increase in expenditures for research and
development in energy."[53]

Beyond the general optimism within government regarding scientific instruments for problem solving was the attractiveness of programs that could be formulated as a public good. Research and development was amenable to this characterization, and the programmatic
expression of this set of initiatives involved in essence the channeling
of money into private-sector projects. R & D funding thus became
attractive to successive administrations because it was an area where
the state could dominate the debate. The "political function of the
science policy apparatus" has generally been interpreted as a strategic

[52]James Everett Katz, *Presidential Politics and Science Policy* (New York: Praeger,
1978), p. 233; see also Barfield, *Science Policy.*
[53]"Science, Technology, and the Economy," Hearings before the Subcommittee on
Science, Research and Development of the Committee on Science and Astronautics,
U.S. House, 92d Cong., 2d sess., April 11, 1972, p. 12.

government tool with which officials can rescue a measure of coherence over an otherwise disparate and fractious arena of public policy.[54]

The emphasis on R & D became particularly compelling for U.S. planners and politicians because energy research and development did not encroach on established private interests yet could be used to insinuate a measure of government control over energy investments. In a factious polity, R & D funding generated few opponents; it could be understood as a public good. One analyst notes: "Energy R & D is expensive, but not enough so to hurt the average taxpayer noticeably when it is financed through the massive Federal budget. And who can oppose research on new energy sources? . . . Nor does R & D by the Government necessarily mean eventual control of the technology that results by the Government or other insensitive institutions. Thus it has not been politically difficult for Congress to be generously supportive of virtually every valid energy R & D initiative, and some of less demonstrated validity."[55]

At the same time the actual allocation of research funds tends to be reserved for administrative specialists, allowing state officials a measure of discretion with which to set priorities. In this way R & D spending is particularly attractive for an administratively weak state. Because it implies funding for prospective industries, not for established producers, it allows greater state direction without directly confronting private investment decision making. Department of Energy officials have indicated that research investment funding tended to be directed into parts of the energy industry that traditionally underinvested. For example, government specialists favored projects in coal, such as coal gasification research, because the industry tended to be less developed technologically. Indeed, this logic was foreshadowed by ERDA's administrator, Seamans, at his confirmation hearings. "In the coal industry today . . . there is very limited research and development capability and I am not going to blame any of those who were operating these coal companies—but the truth of the matter is they have not brought along the science and the technology in anything like the imaginative way that the oil companies have in their explora-

[54]See David Dickerson and David Noble, "By Force of Reason: The Politics of Science and Technology Policy," in Ferguson and Rogers, *Hidden Election*, p. 264.

[55]See J. Herbert Holloman and Michael Grenon, *Energy Research and Development*, A Report to the Energy Policy Project of the Ford Foundation (Cambridge, Mass.: Ballinger, 1975), pp. 16–18. The analyst quoted is John W. Jimison, Congressional Research Service, "Energy—Is There a Policy to Fit the Crisis?" Prepared for the use of the Subcommittee on Energy and Power, Committee on Interstate and Foreign Commerce, U.S. House, September 1980.

tion."[56] Government planners sought out similar suboptimal patterns of investment by private industry.[57] But this logic—with government planners finding their "comparative advantage" in relation to industry—also held for relations within government. Because actual funding targets tended to be technical issues, the administrative planners were given opportunities in research and technology to exercise an influence disproportionate to their actual standing in the policy process. In responding to energy challenges, as a consequence, there emerged a pattern of fiscal initiatives which is understandable in terms of a larger set of institutional relations and capacities.

[56]"Nomination of Dr. Robert C. Seamans, Jr., to Be Administrator, Energy Research and Development Administration," Joint Hearings before the Joint Committee on Atomic Energy and the Committee on Interior and Insular Affairs, U.S. Senate, 93d Cong., 2d sess., December 11, 1974, p. 73.
[57]Interviews, Department of Energy, March 1983.

CHAPTER SEVEN

The Market as State Strategy

> There was nothing natural about *laissez-faire;* free markets could never have come into being merely by allowing things to take their course. . . *Laissez-faire* itself was enforced by the state. . . Thus even those who wished most ardently to free the state from all unnecessary duties, and whose whole philosophy demanded the restriction of state activities, could not but entrust the self-same state with the new powers, organs, and instruments required for the establishment of *laissez-faire.*
>
> Karl Polanyi, 1944

American executive officials tried to fashion a workable policy on energy adjustment by resort to several different tools of government, but in each case they confronted constraints. The initial strategy, international adjustment, failed. Emphasis then shifted toward domestic initiatives, but efforts to redefine and extend the role of the federal government in energy production also failed. Executive officials were able to implement a spending program for energy R&D, but this funding did not produce the immediate payoff necessary to alter prevailing patterns of consumption and production. Oil imports continued to climb throughout the mid-1970s. In the absence of success elsewhere, the issue of oil pricing moved to the center of the policy process. Beginning in 1975, but especially after 1978, executive officials came to embrace the decontrol of domestic oil prices as the single most important tool with which to address the problem of energy adjustment.

For most industrial nations, the OPEC price shocks of 1973–74 passed directly through the economy and forced immediate adjustment.[1] The U.S. economy, however, at the moment of the Arab oil

[1] See Edward N. Krapels, *Price Petroleum Products: Strategies of Eleven Industrial Nations* (New York: McGraw-Hill, 1982).

embargo, was under the guidance of Richard Nixon's 1971 wage and price controls. These controls acted to regulate the market and as a consequence effectively shielded domestic consumers from rising world oil prices. When the controls lapsed in 1973, Congress extended oil price controls in emergency legislation to cope with the crisis. Although enacted as "temporary" measures, these regulatory controls lasted for the better part of a decade and were periodically reformulated in ever more complicated programs.[2]

Price controls made the nation's petroleum markets unlikely to respond to any of the government's energy goals: domestic production was discouraged, consumption was indirectly subsidized, and the effective limit on imported oil did not allow for domestic shortages to be made up with foreign supplies.[3] Oil pricing necessarily became politicized. Consumers tenaciously supported the price controls while the large oil companies inveighed against them. At the same time the Ford administration (and later the Carter administration) began in 1975 to seek various ways to dismantle, circumvent, or abridge the controls. Finally, in the spring of 1979, with new executive authority over oil pricing, the Carter administration announced a program of phased oil decontrol.

With the decontrol decision of April 1979, American executive officials finally found a workable solution to the problems of energy adjustment. Unable to develop new interventionist means to stimulate production and encourage conservation, government officials gradually came to embrace market pricing as the most effective method within their reach to advance their goals both in general energy policy and in foreign policy. Understanding now that the market could be used as a tool of national policy, executive officials became involved in efforts to dismantle the regulatory apparatus. The strategic withdrawal of state involvement in pricing and allocation became the most efficacious act of public policy.

The dominant account of oil-pricing policy focuses on the play of interest groups, presenting the history of oil price regulation and deregulation as the active involvement of interest group and congressional politics.[4] This interest-group explanation traces the persistence

[2]These regulatory policies are described in Joseph P. Kalt, *The Economics and Politics of Oil Price Regulation* (Cambridge: MIT Press, 1981).

[3]See Douglas R. Bohi and Milton Russell, *Limiting Oil Imports: An Economic History and Analysis* (Baltimore: Johns Hopkins University Press, 1978), p. 218.

[4]See, for example, David Glasner, *Politics, Prices and Petroleum: The Political Economy of Energy* (Cambridge, Mass.: Ballinger, 1985).

of regulatory controls and their eventual dismantlement to the efficacy of consumer and industry interests within the policy process. Shifts in world oil markets and the domestic political environment were followed by shifts in the bargaining position of these groups, and change in public policy was the consequence.

Consumer and industry groups certainly did have substantial interests at stake in oil pricing policy, but an exclusive focus on interest groups cannot satisfactorily explain the rise and decline of oil price controls. Societal interests were strongly divided over the merits of decontrol and had reached stalemate.

The explanation I advance here pays attention to the inputs of executive officials and to the institutional structure of the American state. The particularities of American government shaped the terrain for policy struggle and provided resources with which executive officials could influence policy outcomes. Adequate explanation of the timing and content of oil decontrol must attend to the institutional structure of the state, I shall argue, because that structure provided a basis for executive officials with their own agendas to respond to and influence the evolution of oil pricing policy.

Constrained within a fragmented and decentralized government, executive officials were nonetheless able to marshal resources uniquely available to the state. Of particular importance was the special access executive officials had to the international system. This privileged position was obvious in 1978 at the annual seven-power economic summit, where administration officials maneuvered to tie an American pledge on oil decontrol to German and Japanese agreements to reflate their economies. American regulation of domestic energy prices put upward price pressure on international markets, much to the displeasure of other industrial importing nations. Price controls also encouraged imports indirectly and by 1978 had weakened an already sagging American dollar. Oil pricing policy, therefore, came to impinge on other important interests in foreign economic policy, giving executive officials opportunities to recast the issue of oil pricing policy—to recast what was at stake and what could be accomplished.

It is in this context that an appreciation of the structure of the state is important for an understanding of the development of policy. The state's institutional structure provided resources with which, and sites from which, various factions, including government officials themselves, pursued their interests. Executive officials were able to draw upon the state's unique role as the authoritative agent of foreign policy and thereby recast the politics of oil pricing.

SOCIETAL INTERESTS AND OIL PRICE POLICY

Throughout the 1970s large and vocal groups arrayed themselves on all sides of the decontrol issue. Each policy alternative, ranging from the continuation of controls to their immediate dismantlement, would give advantages to identifiable social groups. Yet an exercise that simply matches policy with interests does not adequately explain policy. It is important, therefore, to take a close look at those interests and their role in the course of policy development. Oil pricing policy is not comprehensible as the straightforward translation of private interests into government action. Those interests were mediated by the structure of the state, a structure that allowed executive officials to shape and influence the decision to decontrol.

Regulation and the Entrenchment of Interests

In the decades prior to the 1973 embargo, petroleum producers had generally had their own way on oil pricing policy. The 1959 Oil Import Program, in place until 1971, protected domestic producers from less expensive imports from the Middle East. Throughout the 1960s these import controls kept domestic oil prices at roughly 30 percent above the world market price. Petroleum policy was designed primarily to stabilize and protect the national petroleum industry from its own and foreign surpluses.[5] In the early 1970s, however, fundamental shifts in world petroleum production and pricing interacted with domestic regulations to alter the politics of petroleum policy. Consumer interests, represented in Congress, became a substantial counterweight to industry interests. With this political transformation came new opportunities for executive officials to fashion a distinctive position on oil pricing.

Rising world oil prices and declining domestic production after 1969 put stress on import controls. Although a liberalization of petroleum markets seemed prudent, the broader problem of inflation led in 1971 to the first phase of wage and price controls. The action reinvented the regulation of petroleum.[6] These wage and price con-

[5]For a discussion of the import program see William J. Barber, "The Eisenhower Energy Policy: Reluctant Intervention," in Craufurd D. Goodwin, ed., *Energy Policy in Perspective: Today's Problems, Yesterday's Solutions* (Washington, D.C.: Brookings, 1981), pp. 229–261. A more general discussion is Gerald D. Nash, *United States Oil Policy 1890–1964: Business and Government in Twentieth Century America* (Pittsburgh: University of Pittsburgh Press, 1968).

[6]Authorization for this action was found in the Economic Stabilization Act of 1970, Public Law No. 91-379. On the wage and price control program see Craufurd D.

trols, reformulated through four phases, coincided with fundamental change in international oil markets. Most important, the world supply of crude oil and refined products began to tighten, thus putting upward pressure on prices, and devaluation of the dollar also contributed to higher import prices.[7] By the beginning of 1973 world oil prices, for the first time in the postwar period, had risen above domestic prices.

These dislocations mirrored similar changes within the domestic oil and refining industry and prompted demands for additional government regulation. During the earlier period of cheaper foreign sources of oil, the import quota had provided a regulatory basis for small, independent refiners to buy cheap oil and flourish by marketing petroleum domestically at cut-rate prices. As domestic and international price differentials reversed and the quota system broke down, these refiners were forced to compete with the majors. Price controls began to have divergent effects on refining and marketing firms, creating demands that the federal government intervene to allocate supplies for purposes of equity.[8] The changing differential between prices for domestic crude and imported crude is summarized in Table 7.

Further pressure for regulated allocation came with the fourth phase of price controls. Introduced in August 1973 in an attempt to encourage domestic production, Phase Four controls separated oil prices into two tiers. Crude oil from existing wells—"old" oil—would continue to be controlled. But "new" oil—defined as supplies that exceeded 1972 production levels—was allowed to rise to world price levels. This arrangement created new problems for many domestic refiners and marketers, because of the differences among them regarding access to the cheaper, controlled oil. Refiners and marketers dependent on imported oil (primarily in the Northeast) had higher costs, and this competitive disadvantage dislocated the market. Re-

---

Goodwin, ed., *Exhortation and Controls: The Search for a Wage Price Policy, 1945–71* (Washington, D.C.: Brookings, 1975); M. Kosters, *Controls and Inflation: The Economic Stabilization Program in Retrospect* (Washington, D.C.: American Enterprise Institute, 1975); J. Pohlman, *Economics of Wage and Price Controls* (Columbus, Ohio: Grid, Inc., 1972); G. Haberler, *Incomes Policy and Inflation* (Washington, D.C.: American Enterprise Institute, 1971).

[7]See William A. Johnson, "The Impact of Price Controls on the Oil Industry: How to Worsen an Energy Crisis," in Gary Eppen, ed., *Energy: The Policy Issues* (Chicago: University of Chicago Press, 1974), p. 104.

[8]Ibid., p. 106; William C. Lane, Jr., "The Mandatory Petroleum Price and Allocation Regulations: A History and Analysis," report prepared for the American Petroleum Institute, May 5, 1981, pp. 23–24. See also C. Owens, "History of Petroleum Price Controls," in Department of Treasury, *Historical Working Papers on the Economic Stabilization Program* (Washington, D.C.: GPO, 1974), pp. 1253–54.

Table 7. Prices for domestic and
imported crude oil, 1968–82
(cost per barrel in current dollars)

	Domestic	Imported
1968	3.21	2.90
1969	3.37	2.80
1970	3.46	2.96
1971	3.68	3.17
1972	3.67	3.22
1973	4.17	4.08
1974	7.18	12.52
1975	8.39	13.93
1976	8.84	13.48
1977	9.55	14.53
1978	10.61	14.57
1979	14.27	21.67
1980	24.23	33.89
1981	34.33	37.05
1982	31.22	33.55

NOTE. Prices are figured in terms of
refiner acquisition costs.
SOURCE. Energy Information Administration, *Annual Energy Review, 1984*
(Washington, D.C., April 1985), p. 123.

gional price disparities and shortages prompted a new interest among
industry and government officials in reform. Thus the price control
scheme was extended to include a mandatory allocation program.[9]

Before 1973, when the threats to national oil producers involved
declining world prices for petroleum, the industry had been able to
enlist government help in protecting prices. The price shocks of the
1970s, however, involved higher world prices, and government policy
had to respond to a larger range of societal interests. Price controls,
moreover, had an unintended effect, strengthening the position of
consumer groups and others who had a stake in the continuation of
those controls. Whereas the history of oil pricing policy would lead
one to expect policy makers to respond primarily to the demands of
producers, price and allocation controls were actually extended and
reworked by Congress to insulate consumers from higher world
prices. In the process, domestic producers were denied the full gains
made possible by OPEC pricing. "The interests of oil users," as one

[9]See Johnson, "The Impact of Oil Prices," pp. 109–10; also Bohi and Russell, *Limiting Oil Imports*, pp. 225–26.

analyst notes, "had superceded those of oil producers in determining the direction and nature of government policies."[10]

The interest of consumers in the maintenance of controls was given voice in Congress. Although officials in the executive branch began to argue for decontrol in 1975, Congress repeatedly voted to extend controls. The costs to consumers would be onerous; when coupled with an excise tax, the burden of decontrol was estimated in 1975 at approximately $24 billion.[11] Many in Congress also believed decontrol would benefit only the large oil producers and would hand over control of domestic pricing to OPEC. The opposition to decontrol was manifest in the Energy Policy and Conservation Act (EPCA) of 1975, which rolled back the price of domestic oil and extended controls for another forty months. At the end of that period the president would need to take positive action, subject to congressional review, to end controls.[12]

An explanation of the persistence of price controls does require a focus on societal interests, therefore, and in particular on those of consumers represented in Congress. Price controls actually had the unintended effect of diminishing the importance of the majors in the policy process; in contrast to the pre-1973 period, oil producers were not at the center of oil pricing policy. Regulatory controls enlarged the political struggle over pricing policy, and consumer interests demonstrated an ability to overwhelm the interests of oil producers and thwart, for a time, the designs of executive officials. Oil producers did gain from the subsequent decision to decontrol, but there is reason to be skeptical about their influence over pricing policy.

The Oil Industry and Decontrol

An interest-group explanation of the shift from controls to decontrol would look to the changing influence of consumer and industry groups in the policy process. In particular, one would expect to find an erosion of consumer interests in Congress, on the one hand, and a better-organized and more active oil industry, on the other. Indeed, the absence of strong influence by producers over pricing

[10]Arthur W. Wright, "The Case of the United States: Energy as a Political Good," *Journal of Comparative Economics* 2 (1978), 171.
[11]Testimony of Frank G. Zarb, in U.S. Congress, House, *Hearings on the Presidential Energy Program*, pp. 20–21. See also Richard H. K. Vietor, *Energy Policy in America since 1945: A Study of Business-Government Relations* (London: Cambridge University Press, 1984), p. 249.
[12]Public Law 94-163, December 22, 1975.

policy throughout the period of controls suggests that the rise in industry influence would need to have been quite remarkable. In fact, however, industry interests were not well organized during the period of controls, and the controls themselves split the petroleum industry by creating winners and losers. An explanation for policy change has to look beyond industry interests.

The regulatory controls redistributed very large amounts of income. The major losers were the domestic producers of crude oil; petroleum refiners and consumers were the winners. Joseph Kalt estimates that if these producers had been able to sell their production at unregulated prices, between 1975 and 1980 they would have enlarged their annual income in amounts ranging from $14 billion to $49 billion (in 1980 dollars).[13] The rents that would have gone to crude producers were instead captured by petroleum users, both domestic refiners and final users of petroleum products. (For a calculation of the redistributive effects of price and allocation controls, see Table 8.)

With prices for domestic crude oil controlled, competition developed among users for access to the cheaper supplies. Government regulations initially allowed access to this oil to be based on existing supply contracts. The result was the transfer of billions of dollars to some domestic refiners with access to price-controlled oil who could refine and sell their products at world prices. The Entitlements Program, enacted in November 1974, spread the subsidy to all domestic refiners. Price controls also created distortions in the downstream use of refined petroleum products, prompting demands for "priority access" to supplies. In effect, the growth of federal regulations for allocation expanded the range of refiner and consumer groups that benefited from the control of prices.[14]

Thus price and allocation regulations fragmented the interests of the American oil industry. Kalt notes: "Even among the largest integrated companies, the effect of federal policy was disparate. As the balance of operations shifted from domestic crude oil production (where regulatory burdens were imposed) to refining and international operations (where entitlements benefits were conferred), companies acquired vested interests in the overall regulatory program—

[13]Kalt, *Economics and Politics*, pp. 213–21.
[14]See Joseph Kalt, "The Creation, Growth, and Entrenchment of Special Interests in Oil Price Policy," in Roger G. Noll and Bruce M. Owen, eds., *The Political Economy of Deregulation: Interest Groups in the Regulatory Process* (Washington, D.C.: American Enterprise Institute, 1983), p. 106.

Table 8. Estimated distributional effects of petroleum price controls and entitlements in the United States, 1975–1980 (in U.S. $ billions, 1980 dollars)

	Crude oil producers	Petroleum refiners	Petroleum product consumers
1975	−23.9	+15.0	+6.9
1976	−18.9	+10.2	+6.8
1977	−18.7	+10.4	+6.4
1978	−14.3	+8.5	+4.7
1979	−32.6	+21.8	+8.3
1980	−49.6	+31.7	+12.2

NOTE. 1980 figures are annualized from data for January–March.

SOURCE. Joseph Kalt, *The Economics and Politics of Oil Price Regulation* (Cambridge: MIT Press, 1981), p. 216.

as differences in companies' lobbying efforts on the issue of decontrol repeatedly testified."[15]

Regulatory controls provided gains to some and losses to others within the petroleum industry, and company positions varied accordingly. Standard Oil of Ohio and Marathon, both medium-sized oil companies, are illustrative. Sohio, with a large refining capacity and little oil production of its own, qualified as an independent refiner. Under the terms of the Allocation Act it was able to buy subsidized crude from other domestic producers and consequently favored extension of the regulatory program. Marathon, on the other hand, was largely self-sufficient in petroleum production and was forced to sell some of its production. Unable to gain the full rents of its own petroleum production, Marathon vigorously opposed controls.[16]

The problem with any society-centered explanation of decontrol is that group interests were highly mediated by the prevailing institutional structures of government. The capabilities of societal actors cannot be taken as given. The interests of consumers, which otherwise would have been diffuse and difficult to organize, were crystallized in unanticipated ways by the existing price controls, magnifying consumer claims and creating an effective counterweight to the interests of oil producers. At the same time the regulatory program also split the petroleum industry on the issue of decontrol. Although most oil

[15]Ibid., p. 109.
[16]Vietor, *Energy Policy in America*, p. 247.

173

producers had a substantial stake in decontrol, the evolution of policy in that direction cannot be explained by changes in their capabilities. Indeed, oil decontrol was accomplished in spite of rather than because of the interests of the majors. The shift from controls to decontrol hinges on the role of executive officials who drew upon resources available only to them and, by so doing, redefined the issues at stake.

State Structure and Policy Change

Federal petroleum regulations generated profits and losses to various societal groups, but they also had aggregate effects on the national economy and the nation's international position. Artificially low domestic prices created incentives to overconsume imported crude and to underproduce domestic reserves. Regulations put upward pressure on the world market price of oil, strengthening the monopolistic pricing of OPEC and disadvantaging other industrial nations. Domestic controls prevented the production of domestic oil that, even at world prices, would have cost less than the imported oil that replaced it. The national wealth transferred to foreign suppliers by virtue of control has been estimated at between \$1 and \$5 billion (in 1982 dollars) per year between 1975 and 1980.[17] Executive officials began in 1975 self-consciously to address these national and international consequences of domestic regulation. It was the activity of these state officials which shifted the balance of forces in favor of decontrol, a process accomplished by redefining what was at stake in oil pricing policy.

I have argued that in explaining adjustment policy, we do well to pay attention to the institutional configuration of government and the influence of evolving political institutions on the political position of and resources available to specific actors in the policy process, including executive officials. In this sense, the state is not a single, integrated, institutional actor but a piece of strategically important terrain, which shapes the entire course of political battles and sometimes provides the resources and advantages necessary to win them. Among these resources is the special access of executive officials to the international system.

A focus on state structures allows us to appreciate the independent

---

[17]Kalt, "Creation, Growth, and Entrenchment," p. 101. On international market effects of domestic petroleum regulations see Edward Fried and Charles Schultze, eds., *Higher Oil Prices and the World Economy: The Adjustment Problem* (Washington, D.C.: Brookings, 1975), p. 67.

and intervening efforts of executive officials to recast the problem of oil pricing in ways that advanced a set of goals larger than those embraced by any particular social group. Although decontrol did provide benefits to national oil producers, it was the task of executive officials to redefine what goals the market pricing of petroleum would serve. The need to redress losses to national wealth and to the international position of the nation required state officials to distance themselves from the interests of the oil producers. Their goals, as Douglas Bohi and Milton Russell argue, were "clouded because of pervasive suspicion of the oil industry and because this national goal coincided so exactly with certain of its special interests."[18] The development of a foreign policy rationale for decontrol, most prominently expressed at the 1978 Bonn summit, allowed that decision ultimately to be made.

THE STATE AND MARKET STRATEGY

During both the Ford and early Carter administrations, executive officials articulated a "state interest" in market pricing and related their goals to larger national economic policy and to foreign policy. Both administrations labored under the regulatory programs championed in Congress. In 1978, as we shall see, new opportunities arose for executive action on oil pricing—opportunities seized upon by senior officials and linked to larger objectives in foreign economic policy.

The Ford Administration and Market Pricing

In 1975, although economic recession had restrained growth in oil consumption, the import dependence of the United States continued to climb.[19] In this context the new Ford administration began to articulate a national adjustment strategy organized around a return to the market pricing of petroleum. This priority, which was to re-emerge in the Carter administration, had an uneasy history. At each turn the Ford administration was forced to compromise with Congress, diluting its market strategy. Nonetheless, a small victory was won: congressional legislation contained the provision for eventual presidential control over oil pricing policy.

[18]Bohi and Russell, *Limiting Oil Imports.*
[19]Oil imports from Arab OPEC members constituted the greatest increase. Department of Energy, *1982 Annual Energy Review* (Washington, D.C.: GPO, 1983), p. 57.

The administration made a commitment to decontrol in January 1975 with the proposal for an excise tax and import fee on foreign crude oil and a "presidential initiative to decontrol the price of domestic crude oil."[20] The rationale for decontrol was simple. The deputy administrator for the Federal Energy Administration argued before a congressional committee in August 1975: "The most efficient way to reduce demand and increase supplies (and thereby reduce imports) is, of course, through the price mechanism." The controls frustrated the goal of lowering U.S. dependence on imported oil, and they had other problems as well. The administrator argued that the controls hindered competition in the petroleum industry and prevented rational decision making on corporate investments. He also argued that the controls prolonged distortions and inefficiencies in the adjustment process itself. "As domestic production continues to decline at differing rates in different parts of the country, necessary adjustments in crude oil distribution channels cannot be resolved through the operation of normal market mechanisms, and can only be accomplished [under present circumstances] by ad hoc action by the FEA, which is ill-equiped to deal with such matters."[21]

In addition to this immediate rationale for decontrol were other arguments concerning foreign policy goals. The State Department took the position that leadership in the West hinged on gaining control of rising petroleum imports. International energy goals would necessarily involve domestic pricing decisions to encourage conservation and the development of alternative energy technologies.[22] This indirect support for reformed pricing, based on broad foreign policy goals of allied energy cooperation, were espoused by the Carter administration and became an important force behind the final decision to decontrol.

With this rationale for decontrol, the administration prepared itself for legislative maneuverings with a skeptical Congress. An import fee, proposed in Gerald Ford's January 1975 speech, was designed as a bargaining chip. Ford intended to decontrol oil prices on domestic oil

[20]State of the Union Message, January 13, 1975. The price deregulation proposals made by the administration were contained in S. 594/H.R. 2650.

[21]John A. Hill, in "Oil Price Decontrol," Hearings before the Commitee on Interior and Insular Affairs, U.S. Senate, 94th Cong., 1st sess., September 4 and 5, 1975, pp. 147, 149–50.

[22]A report at the time noted: "As formulated by Secretary of State Henry A. Kissinger, approved by principal officials of the Administration and accepted by President Ford, the high price concept lies at the heart of the government's foreign policy strategy in the energy field." *National Journal*, March 8, 1975, p. 357.

on April 1, 1975, which he could do if Congress did not veto decontrol (as it could under the terms of the 1974 Emergency Petroleum Allocation Act). He was therefore eager for Congress to approve his excise tax proposals and asked for passage within three months. To stimulate congressional passage, Ford warned that he would impose a dollar-a-barrel import fee on February 1 and a second dollar-a-barrel fee on April 1.

This pressure on Congress was not, however, effective. To begin with, Congress passed legislation denying the president the authority to impose an import fee for ninety days. Ford quickly vetoed this legislation. Negotiations followed between the two branches, and Ford agreed to delay both the fee and the decontrol plan. Decontrol action was pushed back to May 1, which was also the date by which Ford wanted Congress to act on his tax proposals. The deadline was missed again, and on April 30 Ford agreed to a phased program of decontrol.[23]

The president could have allowed prices to be decontrolled had he let the Emergency Petroleum Allocation Act expire as planned on August 31, 1975. Ford, however, wanted decontrol to be a decision shared with Congress, and so he accepted the phased plan.[24] A second opportunity to veto energy legislation that extended EPAA controls came in December 1975 (the Allocation Act was due to end on the fifteenth of that month). Again the president chose not to decontrol by veto, waiting instead for the phased program to lapse in thirty-nine months. Thus the president, although urging decontrol, was not willing to take it at any cost.

A victory of sorts was rescued from the congressional struggle. The Energy Policy and Conservation Act of 1975 contained provisions for eventual presidential discretion over oil pricing policy. Although this authority was still several years off, the politics of oil pricing had been shifted from a congressional-executive struggle to an intra-administration struggle. In the early Carter administration, officials were still bound by EPCA controls, and they sought an ingenious taxing scheme to replicate market pricing artificially. As discretionary authority over pricing policy neared in 1978, the struggle moved into the executive branch, where foreign policy and energy officials sought to fashion a persuasive rationale that would lift decontrol above the troubling constraints of domestic politics.

[23]Congressional Quarterly, *Energy Policy*, 1981, pp. 34–35.
[24]Ibid., pp. 35–36.

Market Pricing by Artifice

The Carter administration came to office officially opposing decontrol of petroleum prices, but it came to embrace a taxing scheme that moved effective prices to world levels. Once in office, energy planners moved quickly to propose a tax scheme that would bring domestic energy prices to the prevailing international level—the Crude Oil Equalization Tax (COET) proposal of 1977. This elaborate tax proposal languished in Congress and was ultimately defeated. When energy prices moved upward again with the collapse of Iranian oil production, the Carter administration came to embrace the strategy of market adjustment more fully.

Jimmy Carter's initial pricing proposals were included in the policy package presented to Congress as the 1977 National Energy Plan. The oil price taxing aspects of the plan reflected the tactical problems of using the price mechanism to alter consumption patterns while also meeting obdurate political resistance in Congress (primarily among liberal Democrats in the House) against the idea of giving producers a free market in petroleum. The 1977 plan's Crude Oil Equalization Tax was designed as a tax on producers in an effort to limit their profits while at the same time bringing the effective price of oil at the refining and marketing stages up to world levels. Taxes collected from oil producers were to be rebated back to lower- and middle-income families in order to offset the inflationary impact on these households of higher retail energy prices. Tax receipts would also be channeled into mass transit projects and an energy investment fund.

In making prices to consumers reflect international prices the COET proposal addressed only the consumption side of the problem. The tax would not create new production incentives. As Mark Steitz notes, "one possible method for dealing with the consumer-to-producer transfer had been previously proposed; President Carter's crude-oil equalization tax would have prevented any of the revenue gain from decontrol from accruing to producers. While this would have maintained the demand-side benefits of decontrol, the supply-side benefits would have been lost; no additional domestic production would occur."[25]

This emphasis on conservation rather than production allowed the National Energy Plan to employ a tax in order to dampen consumption by moving prices toward replacement cost. Carter's policy inno-

[25]Mark Steitz, "Oil Decontrol and the Windfall Profits Tax," in Raymond C. Scheppach and Everett M. Ehrlich, eds., *Energy-Policy Analysis and Congressional Action* (Lexington, Mass.: Lexington Books, 1982), p. 82.

vation was the attempt to have it both ways: to continue controls in order to regulate the prices producers could charge, but also to introduce replacement-cost pricing for energy. With this initiative the administration affirmed that price levels had to be set in a larger context of demand and supply. Elaborate though it was, the new notion of pricing represented a step away from the premises of the regulatory regime.

This distinctive approach was slowly smothered in congressional committee. In the House, where Democratic leadership supported the package, the COET proposal survived committee and floor votes. The Senate, however, was less forthcoming, and the decisive obstacle was the Finance Committee. Although finance chairman Russell Long from Louisiana initially endorsed COET, the committee refused to pass the tax proposal, and it was hung up in Congress without legislative life.

THE TRANSNATIONAL BARGAIN

In early 1978 the Carter administration was caught between an annual import bill for oil of $45 billion and a failed proposal for a COET. At the same time leaders of other industrial nations began to voice criticisms of excessive American consumption of oil. The administration's foreign economic policy officials then began an effort to redefine the oil pricing issue by linking oil decontrol to a larger set of international economic problems and to a diplomatic agreement that addressed those problems. In doing so, they drew on the special legitimacy of the state to manage foreign economic policy. In transforming the issue of decontrol, these officials strengthened their position within the domestic policy process and within the administration itself. Less than a year later, the administration had successfully begun the process of decontrol.

Although leaders from the major industrial countries had become highly critical of rising American oil consumption, in 1978 U.S. officials were more concerned about slow economic growth in Japan and West Germany. Using the annual summit conference as their forum, they attempted to forge a transnational bargain linking these issues. Allied governments, Germany most prominently, were attempting to pressure the American government to bring the cost of domestic oil into accord with prevailing world prices. Meanwhile, the Carter administration was attempting to convince Chancellor Helmut Schmidt to reflate the German economy in order to stimulate international growth. It was this set of counterdemands which made the decontrol decision of larger significance and gave a new rationale to administration officials who already favored market pricing.[26]

In the winter and spring prior to the 1978 Bonn summit, the industrial nations confronted several major international economic issues. The German and Japanese economies were experiencing both low inflation and low growth. The United States, on the other hand, had accelerating inflation, an increasing oil deficit, and a weakened currency. The Carter administration moved to reflate the U.S. economy and sought similar efforts from Germany and Japan. Also, the Multilateral Trade Negotiations were entering their final stages, and the British and French governments were showing some reluctance to finish the process. Finally, there was the nagging issue of American energy consumption—by the end of 1977 oil imports had risen to a historic high. A package agreement might be possible.

It was probably British prime minister James Callaghan who in the early months of 1978 began to explore a package deal for the Bonn summit. An agreement linking the issues was discussed when Callaghan visited the White House in the winter. Carter's key officials for international economic policy—Richard Cooper at State, Anthony Solomon at Treasury, and Henry Owen, ambassador at large for economic summits—had also discussed the idea at the first of the year.[27] Formal discussions to link American energy policy and German growth policy began in March 1978 at meetings of the summit sherpas in Bonn.[28]

In Bonn the American representative, Henry Owen (special representative for summit preparations) met with Chancellor Helmut Schmidt seeking a commitment on German reflation. The German chancellor asked for American concessions, contending that the continuation of controls on American domestic petroleum indirectly contributed to higher world oil prices. Not only did controls discourage the production of domestic American oil, they also encouraged consumption and therefore increased demand in the international mar-

---

[26]The most complete description of this fascinating bargain is in Robert D. Putnam and C. Randall Henning, "The Bonn Summit of 1978: How Does International Economic Policy Coordination Actually Work?" Paper presented at the workshop Intergovernmental Consultations and Cooperation about Macroeconomic Policies, Brookings Institution (Washington, D.C., April 16, 1985). See also George de Menil and Anthony M. Solomon, *Economic Summitry* (New York: Council on Foreign Relations, 1983); Robert D. Putnam and Nicholas Bayne, *Hanging Together: The Seven-Power Summits* (Cambridge: Harvard University Press, 1984).

[27]Interview, Henry Owen, August 10, 1985.

[28]Putnam and Henning, "Bonn Summit of 1978," p. 85.

ket. In exchange for budgetary stimulus in Bonn, Schmidt asked for assurances that the United States would cut levels of imported oil and go forward with its comprehensive energy policy. The result was a trade-off: German reflation for American world oil pricing.

The nature of this bargain was suggested by Chancellor Schmidt just prior to the summit: "Governments of some participating countries believe that they have a recipe for me and for Germany. By way of compromise, if others would bring about some sacrifices or tackle some domestic hardships, I would be ready to do so in my country." Carter also has indicated that a bargain had been struck. In his memoirs he notes that he held a meeting with congressional leaders several weeks before the Bonn summit. "I got all of those who would speak out to advise me . . . to tell our partners at the Bonn economic summit meeting that if Congress did not act to raise the domestic price up to the world level by 1980, then I would act administratively."[29]

The agreement at Bonn did not, strictly speaking, involve an exchange of concessions. Most participants suggest the heads of state themselves wanted the German pledge on growth and the American pledge on energy. Rather, the agreement was a form of cooperation that attempted to strengthen each leader's domestic position. Carter and Schmidt would each be able to pursue their policy commitments bolstered politically by the impression that a concession had been extracted from the other. Thus the Bonn pledge was the outcome of a momentary "international coalition" of political leaders, each with domestic political problems, each agreeing to create the convenient fiction that hard-fought concessions had been won.[30] Ironically, at a moment when the interests of the leaders of two powerful industrial

[29]Schmidt is quoted in John Vinocur, "Schmidt Says U.S. Holds Key to Economic Accord," *New York Times,* July 14, 1978, p. 3; Jimmy Carter, *Keeping Faith: Memoirs of a President* (New York: Bantam, 1982), p. 104.

[30]An interpretation along these lines was suggested to me by Richard Cooper, interview, August 9, 1985. The Japanese sherpa to the summit, Deputy Foreign Minister for Economic Affairs Hiromichi Miyazaki, argues that the issues of economic stimulation and energy were not explicitly linked, and that the Japanese, at least, had no intention of "bartering" over U.S. oil pricing policy. The Japanese did resist engaging in a "locomotive" expansion of American, German and Japanese economies. Nonetheless, a compromise did emerge. "The Japanese," Miyazaki noted in an interview, "would do things that would look like reflation in U.S. eyes." In essence, Miyazaki's argument is that each side was able to use the other's public statements in the formulation of a domestic rationale for policy they already supported. In this sense, the summit pledges were used to strengthen the domestic position of the various leaders, and they were not strictly speaking a bargain built on mutual concessions. Interview, Ambassador Hiromichi Miyazaki, Bonn, Federal Republic of Germany, July 22, 1985.

nations converged, agreement had to be presented to the world as a contest over concessions.

There was a great deal of ambiguity within the Carter administration concerning what the president had in fact pledged. Foreign policy officials understood that the Bonn statement was itself a policy decision on oil pricing. One participant suggests that the decision was made "in principle" at Bonn that the effective price of domestic oil was to move to world levels. However, this commitment might still be carried out through the COET or other tax or administrative mechanisms.[31] Another participant argues even more strongly that the decision on decontrol was made *in* Bonn. "Carter wanted oil decontrol," this official argues, "and welcomed Bonn as an opportunity to do it."[32] Carter himself argues that the U.S. inability to implement an energy policy was "becoming an international embarrassment." And he notes his commitment at Bonn to "let American oil prices rise to the world level." Others, such as CEA chairman Charles Schultze, thought Carter had gone too far, exchanging a pledge on oil pricing for commitments on growth that the Germans would have pursued anyway. Schlesinger was hesitant to see the government make a foreign policy commitment when the issue hinged on domestic politics. The domestic policy staff, however, were least aware or not convinced that a decision had in fact been made at the summit. Although they had attended several presummit meetings on the proposed pledge, Stuart Eizenstat and his staff were dismayed by the announcement at Bonn. Eizenstat faulted the foreign policy side of the administration for not being "sensitive to domestic considerations." No decision memorandum had been prepared before the summit, and the domestic implications of the pledge had not been discussed.[33]

A foreign policy rationale now added impetus to government deliberations on oil pricing policy. Although the Bonn summit pledge was pushed by officials not directly involved in domestic policy discussions, it did encourage those discussions. The pledge, one domestic

[31]Interview, Richard Cooper, August 9, 1985.

[32]Interview, Henry Owen, August 10, 1985. In an essay Owen described the outcome of the Bonn summit: "the United States decontrolled oil prices (arguably the single most important step taken by the industrial countries to address the world energy problem), which Carter could only have done as part of an international economic bargain which also included stimulus and trade pledges from other countries." Owen, "Taking Stock of the Seven-Power Summits: Two Views," *International Affairs* 60 (Autumn 1984), 660. It would come as a surprise to other officials in the administration (particularly on the domestic side) that Carter decontrolled oil prices at the Bonn summit.

[33]Carter, *Keeping Faith*, pp. 103–4; interview, James Schlesinger, August 22, 1985; interview, Stuart Eizenstat, June 19, 1985.

analyst noted, "gave people within the administration reason to keep working on this issue."[34] A student of the annual Western economic summit meetings concludes that "international pressure played a significant part in convincing President Carter that the time had come to push for decontrol of the price of oil." One Carter foreign economic policy official suggests the Bonn pledge "tilted the balance" in administrative deliberations on oil pricing.[35] Another notes it would have been "embarrassing for Carter to change his mind after Bonn." A cabinet secretary argues that the Bonn pledge was a "significant but not preponderant" factor in the final decision to decontrol.[36] Nonetheless, in internal discussions of oil pricing and the conflicting problem of inflation, in early 1979, Carter had clearly realized that a postponement of decontrol would be possible only if Schmidt and other Western leaders could be persuaded that inflation posed the greater international economic threat. After Bonn, oil pricing was manifestly an international issue, even in meetings with domestic staff.

The impact of this transnational bargain is difficult to measure. Robert D. Hormats, a State Department official who helped guide American summit planning during the Ford and Carter administrations, has emphasized the summit's importance in countering domestic opposition to decontrol. The summit, he has noted, "helped Carter, who I think wanted to do it in the first place, but for a variety of reasons didn't."[37] In this interpretation the summit did not force the decontrol decision but supported the decision. Other officials see a stronger significance in summit politics. The Bonn pledge served to break the deadlock on oil pricing within the administration, Robert Putnam and Randall Henning argue, by separating the question of whether to raise domestic oil prices from the means for doing so.[38] This idea overstates the role of the Bonn pledge. On at least two occasions in early 1979 Carter had indicated privately to his staff that fighting inflation might be of greater importance than living up to the Bonn pledge and that he would be willing to accept the consequences. The crux of the decision was the dilemma between meeting the Bonn pledge and setting goals for energy conservation and production, on

[34]Interview, Jim Voytko, June 7, 1985.
[35]George de Menil, "From Rambouillet to Versailles," in de Menil and Solomon, *Economic Summitry*, p. 24; interview, Richard Cooper, August 9, 1985.
[36]Interview, Henry Owen, August 10, 1985; interview, James Schlesinger, August 22, 1985.
[37]Quoted in Clyde H. Farnsworth, "Trade Notes," *New York Times*, April 15, 1984, p. 6:8.
[38]Putnam and Henning, "Bonn Summit of 1978," p. 91.

the one hand, and fighting inflation through the continued use of controls, on the other. In the end the weak-dollar problem seems to have undercut the inflation-fighting argument. Treasury Secretary Michael Blumenthal and Anthony Solomon argued that continued controls would further weaken the dollar, thus cancelling out anti-inflation gains.[39]

The transnational bargain struck at Bonn had an agenda-setting effect. It ensnared the administration debate over oil pricing in foreign policy and international economic issues. Energy officials were forced to balance their domestic concerns against international problems. In the end, officials in energy and foreign policy became allies, articulating a rationale for decontrol that challenged officials concerned with domestic politics and the inflation-fighting program.

AGENDA SETTING AND THE DECONTROL DECISION

The Bonn agreement signaled a shift in the way in which the Carter administration characterized oil pricing, thereby changing the balance of forces for and against decontrol. Within the administration, senior officials concerned primarily with national and foreign economic policy, who uniformly favored market pricing, gained the political upper hand. The administration also strengthened its bargaining position with Congress. Price and allocation controls were due to expire in April 1979, giving the president for the first time in almost four years discretionary authority over pricing. A decision would be necessary: he could choose to continue controls, abolish them, or phase them out. The struggle over pricing policy continued, but executive officials pursuing a larger, foreign economic policy agenda had become stronger. It is this development which is necessary to explain policy change.

Circumstances external to administrative strategy pushed a decision on decontrol forward in late 1978 and early 1979. International oil markets were in turmoil. The shutdown of Iranian production triggered skyrocketing prices on the spot market, and in December 1978 and March 1979 OPEC made decisions that doubled world prices. At the same time his Western allies continued to pressure Carter to stand by his 1978 pledge to raise domestic prices to world levels. They had reason to complain. Whereas Japanese and Europeans had reduced import levels marginally, American oil import

[39]Eizenstat diary, January 3 and March 15, 1979; on the weak-dollar problem, Eizenstat diary, January 18, 1979.

Figure 4. Decision memorandum for President Carter on oil pricing, January 3, 1979.

Energy policy
—replacement cost pricing;
—provision of adequate incentives to stimulate maximum domestic production of oil;
—incentives for conservation and a reduction in oil imports;
—equity in the distribution of any windfalls associated with oil price increases;
—elimination of the current complex system of price controls, allocation, and entitlements.

Economic policy
—reducing inflation, and holding 1979 increases in the Consumer Price Index as low as possible;
—maintaining the strongest possible posture to urge major unions to remain within Administration guidelines in upcoming contract negotiations;
—maintaining growth in the economy and in employment;
—improving the balance of trade and the strength of the dollar; and
—regulatory reform objectives.

Foreign policy
—the Bonn pledge to raise the price paid for oil in the U.S. to world prices by the end of 1980
—the general international concern over inflation, including Bonn pledge to make reduction of inflation a top priority of U.S. economic policy.
—reducing U.S. dependence on oil imports, thereby reducing the trade deficit and strengthening the dollar.
—maintaining U.S. credibility among our key Summit allies and assuring their continued efforts toward meeting their own Bonn pledges.

SOURCE. "Memorandum for the President. From Jim Schlesinger, Mike Blumenthal, Richard Cooper, Charles Schultze, Alfred Kahn, Jim McIntyre, Henry Owen, Stu Eizenstat. Subject: Domestic Crude Oil Pricing—Information," January 3, 1979, pp. 2–3.

levels continued high by historic standards. Finally, inflation continued to mount, compounding the seriousness of the oil-pricing decision.

All of Carter's senior advisers for domestic and for foreign policy were engaged in the debate over oil pricing. No longer was decontrol simply an energy issue; national economic and foreign policy considerations were explicitly incorporated in the decision process. The expansion of the policy debate is illustrated in the decision memorandum prepared for the president by his senior advisers (see Figure 4).

The memorandum summarized the key issue before the president: "Should our energy policies and international commitments on energy be deferred or delayed in their implementation so as to minimize the near term inflation effects which an increase in U.S. prices to world levels would entail?"[40] Those urging immediate decontrol in-

[40]"Memorandum for the President," January 3, 1979.

cluded energy secretary Schlesinger, who saw it as a way to provide incentives for production and conservation and reduce imports.[41] Treasury secretary Blumenthal and his deputy, Anthony Solomon, urged decontrol as a means of improving the balance of trade and strengthening the dollar. Richard Cooper and Henry Owen, along with Blumenthal, had been instrumental in the original Bonn pledge, and these officials continued to press for decontrol for foreign policy and international economic reasons.

Those giving special attention to inflation and domestic political considerations, and seeking a compromise position, included Charles Schultze of the CEA and Alfred Kahn, who headed the Council on Wage and Price Stability. Special Trade Representative Robert Strauss and OMB director James McIntrye also sought compromise. Those most resistant to immediate decontrol had an eye on Democratic party politics and constituencies: Stuart Eizenstat and Vice-President Walter Mondale.

In the meantime officials debated various options on production, conservation, inflation, and foreign policy. The critical meeting was at Camp David on March 19. Gathered with his cabinet secretaries, Carter listened to a vigorous debate on oil pricing. Schlesinger presented what had become a narrowed set of options: either immediate decontrol or a gradual raising of prices to world levels. Schlesinger argued that decontrol was necessary to meet the Bonn pledge; that it would be a symbol to the IEA and OPEC of the administration's seriousness about gaining control over imports; and that, in the absence of a presidential decision, Congress would lead on the issue. Vice-President Mondale spoke in favor of continued control, with phrased decontrol and a windfall tax as second-best. Blumenthal argued that immediate decontrol would allow the president to be bold, as he had been with Anwar Sadat and Menachem Begin on the Middle East accord reached at Camp David. Such a decision would help the dollar, and the effect on inflation would be only marginal. Eizenstat argued that in political terms, immediate decontrol would "kill us on inflation." Carter ended the discussion by suggesting he was hesitant to decontrol without a tax—he could not simply give $16 billion to the oil companies. Total decontrol with a tax to allow for some "consumer compensation" was what he favored.[42]

Following this Camp David summit, discussions focused on what

---

[41]Schlesinger thought that a "window of opportunity" had been created by the Iranian crisis in the early months of 1979 that would close by summer. Interview, August 22, 1985.

[42]My source for this paragraph is Eizenstat diary, March 19, 1979.

type of decontrol would provide the most likely basis for congress to pass a windfall profits tax. At a meeting on March 23, 1979, Carter again heard the foreign policy and economic arguments for decontrol. The Council of Economic Advisers, which had wavered on the issue, argued that a phased program could be completed by 1981. Warren Christopher from the State Department argued that decontrol would enhance Carter's image for world leadership by honoring the Bonn and IEA pledge.

On April 5, 1979, Carter announced the United States would gradually lift price controls on domestic crude oil.[43] Existing law would have ended controls in September 1981; Carter phased in decontrol over several months. At one level it was a dramatic turnaround by the Carter administration, as observers noted.[44] But in a more important way it was not. Executive officials had consistently sought policy opportunities to discourage consumption and boost production. Earlier incarnations of energy policy—such as the Crude Oil Equalization Tax—had sought to influence price within prevailing political constraints. The new problems created by the oil shock of 1978–79 also provided new opportunities for extensive action. Indeed, the new policy was seen as a victory for the planners of Carter's first energy plan.[45]

Strikingly, the other groups that had struggled over oil pricing throughout the decade were now on the outside. Congressional opponents of decontrol were on the defensive. Several congressional representatives continued to attempt to discredit the decision by linking it to the interests of the oil industry.[46] But the Congress was no longer able to find a majority vote for continued controls. In the fall of 1979 Representative Toby Moffett of Connecticut proposed to continue price regulations, but for the first time in history the House voted against the extension of controls. The amendment never came to the floor in the Senate.[47] Consumer interests in controls had not changed—indeed, the 1979 oil shock made controls more attractive. What had changed was the terms of the debate.

[43]Martin Tolchin, "Carter to End Price Controls on U.S. Oil and Urge Congress to Tax Any 'Windfall Profits,'" *New York Times*, April 6, 1979, p. 1.

[44]See, e.g., Steven Rattner, "Decontrol a Complete Turnaround in Strategy," *New York Times*, April 6, 1979, p. D3.

[45]Richard Halloran, "A Schlesinger Victory Seen in Energy Plan," *New York Times*, April 7, 1979, p. 38.

[46]Senator John Durkin, speaking for the New England delegation, argued that Secretary of Energy Schlesinger shared "the oil industry's self-serving allusion that all will be well if we only pay higher prices." Quoted in Vietor, *Energy Policy in America*, p. 265.

[47]*Congressional Quarterly*, October 13, 1979, p. 2262.

The oil industry was also on the outside of the decontrol decision. The administration proposed a Windfall Profits Tax designed to capture half of the expected increase in oil revenues. In effect, it applied a severance tax of 50 percent of revenues gained from the release of controlled production. The large oil producers, while opposing the tax, were hurt less by this tax than by the earlier COET program. Independent producers, whose revenues came almost exclusively from domestic reserves, vigorously attacked the new tax. Congress struggled over the tax plan, and old consumer and industry conflicts resurfaced, but the final agreement preserved the general character of the Carter proposal. An estimated $227 billion in tax revenues would be transferred from the oil industry to the federal treasury by 1988.[48]

Neither consumer nor industry interests were satisfied with the outcome, though the final tax and decontrol plans bore their marks. The movement toward market pricing, however, was propelled by a distinct agenda embraced by executive officials. These officials were strengthened as oil pricing became a foreign policy concern, most forcefully defined as such at the 1978 Bonn summit. In a contentious policy process, executive officials were able to resort to resources unique to the state—the ability to gain special access to the international system and to define issues in terms of foreign economic policy imperatives.

CONCLUSION

In an era of turbulent oil prices, the choice between regulatory and market policy engages powerful social and economic interests. For consumer and producer groups the struggle is over who will bear the burden of higher energy costs, and each policy choice produces winners and losers. For executive officials, with a broad political mandate, the problem is not simply the mediation of societal demands. Foreign policy and national economic goals are also at stake. Such were the circumstances in the protracted struggle over oil pricing in the 1970s.

In providing an explanation of oil pricing policy—of the persistence of price controls and their eventual abandonment—I have focused on both societal groups and government officials. The society-centered approach accounts for policy change in terms of shifts in

[48]See Vietor, *Energy Policy in America*, p. 270.

the power or influence of consumer and producer groups that struggle over pricing policy. This focus on interest groups is useful in understanding the persistence of controls, but even in this case those interests were highly mediated by the prevailing structures of government. If we take the capabilities of consumer and industry groups as given, we cannot fully explain the movement toward decontrol. Price controls had the unintended effect of diminishing the importance of petroleum producers in the policy process; in contrast to the pre-1973 period, industry groups were relegated to the margins of pricing policy. Regulatory controls served to enlarge the struggle over policy, and consumer interests represented in Congress demonstrated the ability to perpetuate controls. Although government officials, pursuing their own agenda, were also thwarted during this period, the societal divisions over pricing policy created opportunities for them to advance their own position.

The interest-group explanation accounts for the movement toward market pricing in terms of the rising influence of producer interests. These interests, however, were divided over oil policy. Indeed, regulatory controls helped fragment the petroleum industry by creating winners and losers. The decision to decontrol did not respond directly, or even indirectly, to industry pressure. If executive officials were simply developing policy that echoed the most vocal or widely embraced societal interests, then they would have continued controls as politically far more rewarding. Decontrol policy was actually accomplished in spite of rather than because of industry interests. Executive officials had to differentiate their position on market pricing from the positions of the petroleum industry.

The alternative approach I have developed focuses on the independent impact of executive officials on policy development. In this case, executive officials, concerned about national economic and foreign policy, maneuvered to tie oil pricing to a larger set of international issues and, by so doing, recast what was at stake. In 1979 the executive gained discretionary authority over oil pricing, and by that moment the issue was imbued with larger significance. The new context of foreign and international economic policy tilted the balance in a political setting that otherwise favored the continuation of controls. A foreign policy rationale codified at the 1978 Bonn summit made the decontrol decision possible.

Many groups struggled over oil pricing policy, but an explanation that omits the role of executive officials with their own agenda is incomplete and misleading. The struggle over oil pricing policy unfolded within a distinctive set of government institutions. Those institutions helped shape the decontrol of oil. They did so not in the guise of a single, integrated institutional actor but as a piece of strategically important terrain whose topography influenced the course of battle over pricing policy. First, institutional advantages went to consumers, as price controls unexpectedly sheltered the domestic population from the effects of adverse international economic change. Ultimately, the institutional advantages accrued to executive officials through their special access to the international system. A diplomatic commitment changed the definition of what was at stake in the domestic debate.

CHAPTER EIGHT

Reasons of State

> Men make their own history, but they do not make it just as they please; they do not make it under circumstances chosen by themselves, but under circumstances directly encountered, given and transmitted from the past.
>
> Karl Marx, 1852

In responding to the oil shocks of the 1970s, American government officials were continuously trying to reconcile what was desirable with what was possible. Executive officials sought to encourage energy adjustment in ways that would promote broad goals in national economic and foreign policy. They experienced constraints in doing so, constraints imposed by the institutional character of the American state and by its international and domestic position. To understand the travails of American policy, the search for a workable adjustment policy, and the final triumph of a market approach, we must pay attention to the activities of politicians and civil servants as they attempted to develop and draw upon the state's institutional resources and arrive at a workable policy.

At the center of this inquiry is the issue of state capacity—the ability of the executive officials of the American state to achieve their goals. The notion of the state as an autonomous and weighty actor goes back at least as far as Machiavelli. Indeed, it is at the heart of the Realist vision of international relations. Yet I am concerned here, more so than most Realists, with both the domestic and the international capacities of states. The state's position within these two arenas is linked. The limits on and opportunities for solving adjustment problems in one arena influence the objectives of the state in the other.[1]

[1] Attention to both domestic and international constraints and opportunities on state policy was foreshadowed by the mercantilist writers of the sixteenth, seventeenth, and

The state is not simply an arena of political conflict; it is an institutional structure that gives government officials a unique standing in the pursuit of their own goals and strategic perspectives, which may or may not conflict with those of other actors. From this perspective, the questions for investigation concern the fate of those goals and strategies: How successful are state officials in implementing their policy agenda? How are state goals and strategies modified and reworked in the face of constraints imposed by the institutional structure of the state itself and by the competing goals and interests of domestic and international actors?

In the early decades after World War II, the United States possessed historically unsurpassed international military and economic capabilities. The strength of its international position was manifest across the spectrum of foreign and economic policy issues. During this period the American state could use the nation's extraordinary international power to shape the course of economic change, and this ability lessened the importance of its limited ability to shape domestic adaptation. Difficult domestic decisions on economic adjustment and political struggle over the state's role in the adjustment process could be avoided.[2] The United States could use its power to force others to adjust, or it could use its unparalleled domestic resources and wealth to absorb the impact of adverse international economic change.

In historical terms the oil crises of the 1970s served to end that postwar period. A resort to international strategies of adjustment proved unavailing. Consequently, the American response reveals the effectiveness of the state when international power has waned. When international responses are no longer available, the domestic capabilities of government are thrown into stark relief. American exec-

---

eighteenth centuries. Mercantilist policies were developed to consolidate and expand the powers of the state caught between a fragmented civil society and the universal claims of the church. See Eli F. Hechscher, *Mercantilism*, vol. 1 (London: Allen & Unwin, 1935), p. 21, quoted in Stephen Krasner, "United States Commercial and Monetary Policy: Unraveling the Paradox of External Strength and Internal Weakness," in Peter J. Katzenstein, ed., *Between Power and Plenty: Foreign Economic Policies of Advanced Industrial States* (Madison: University of Wisconsin Press, 1978), p. 52.

[2]This argument is made by Krasner, "U.S. Commercial and Monetary Policy," pp. 52–53. See also Robert Gilpin, *U.S. Power and the Multinational Corporation* (New York: Basic Books, 1975); Charles Maier, "The Politics of Productivity: Foundations of American International Economic Policy after World War Two," in Katzenstein, *Between Power and Plenty;* Fred Block, *The Origins of International Economic Disorder* (Berkeley: University of California Press, 1977); and Alan Wolfe, *America's Impasse: The Rise and Fall of the Politics of Growth* (New York: Pantheon, 1981).

utive officials were forced to confront the limits of those domestic capabilities—probing the scope of state powers, seeking to expand those powers, and ultimately, as I have argued, developing policy within the confines of those powers. If energy adjustment is one case of a broader range of international economic adjustment challenges, then the pattern of response provides a window onto the larger possibilities for American ruling elites to define and implement national policy.

The politics of energy adjustment involved politicians and civil servants in a search for a workable policy that would advantage the nation in the international sphere. When a particular strategy proved difficult to implement, emphasis shifted to other lines of action. These are the microfoundations of state strategy. At the same time the sequence of adjustment policies reveals a logic of government action that is comprehensible in terms of underlying institutional structure. The policy tools and institutional resources available to government officials after 1973 were the precipitates of earlier historical episodes when government and business had competed to influence policy and, in doing so, defined the nature and scope of their respective powers. Macroprocesses of socioeconomic and state structural formation, as well as the outcomes of previous policies, created the "official dispositions" and "state capacities" available in the 1970s. The failure of earlier attempts at institutional transformation limited later possibilities for change. Organizational structures provided the foundation that shaped the capacities and limits of adjustment policy. These are the macrofoundations of state strategy.

Why did the United States follow a circuitous policy course before finally arriving at a policy of market adjustment? The answer is that executive officials came to address energy adjustment in terms of the limited state capacities they had at their disposal, even though initially they had attempted more ambitious interventions. Energy adjustment problems did not arrive on the desks of government officials and politicians prepackaged and sorted. A crisis, after all, continuously provides opportunities to define the problem in new ways. But when officials began the process of problem solving, their struggles to realize policy goals revealed which new channels of action were possible and old ones were necessary. Efforts to enlarge the scope of state capacities were thwarted, and U.S. officials found it necessary to fall back on traditional means to push the adjustment process forward. Ultimately this process involved reconstructing the market for petroleum through regulatory decontrol.

INSTITUTIONS AND THE "SHADOW OF THE PAST"

State officials make their own strategy, but not always as they would choose. A logical, systematic understanding of what states, conceived of as coherent and purposive actors, might want to do when confronted with international economic change is not sufficient to explain what states in fact do. Simplifying assumptions about states as unitary and rational agents are useful for clarifying some types of political phenomenea, but they can obscure others. To understand particular policy outcomes or sequences of strategies, it is useful to adopt a more differentiated, historically grounded, institutional conception of the state. To speak of the state "acting" is actually to confront particular sets of officials, situated in hierarchical and representative institutions, making decisions over time. Existing institutional structures exhibit a logic of their own, as I have argued, and they weigh heavily on the sequence of choices made by officials.

Institutions are established at particular moments of history and tend to persist. The persistence of organizations, and the difficulty of organizational innovation and change, Arthur Stinchcombe explains in terms of the "liability of newness." When circumstances prompt the need for institutional change, he argues, "the newer organization has to be much more beneficial than the old before the flow of benefits compensates for the relative weakness of the newer social structure." For this reason, organizational innovation tends to succeed "only when the alternatives are stark (generally in wartime)." In the absence of crisis the liability of newness makes organizational change difficult. This difficulty stems from several factors. First, new organizations generally involve new roles that must be learned. Specialized skills are developed in established organizations and passed on to new members. New skills and expertise are difficult to develop and must be recruited from outside. The availability of these skills and expertise limits the possibilities for organizational innovation. Second, the process of inventing new roles, the establishment of relations within the new organization, and the structuring of rewards and sanctions have "high costs in time, worry, conflict, and temporary inefficiency." Third, new organizations necessarily involve social relations among strangers. "This means," Stinchcombe argues, "that relations of trust are much more precarious in new than old organizations." Finally, old organizations have a stable set of relations with those who interact with the organization. "Old customers know how to use the services of the organization, have built their own social systems to use the old products or to influence the old type of government, are familiar with the channels of ordering, with performance qualities of the product, with how the price compares, and know the people they have to deal with—whom to call up to get action, for instance."[3] For these reasons, organizational innovation is difficult and tends to take place episodically, when war or other crises overwhelm the value relevant actors derive from predictability and stability.

The institutions within which government officials operate tend to be rigid. Confronted with new policy challenges, politicians and civil servants find it very difficult to redefine their roles and responsibilities—either to expand *or* to contract their powers. Government institutions, as James March and Johan Olsen note, "are relatively invariant in the face of turnover of individuals and relatively resilient to the idiosyncratic preferences and expectations of individuals."[4] Because government institutions are relatively difficult to change, and because some incumbents may have their own institutional interest in maintaining current organizational forms, new socioeconomic challenges will tend to be approached in terms of prevailing policy tools and institutional resources. At any given historical moment, executive officials will be making choices in the context of relatively fixed sets of institutional arrangements. When old institutions shape and influence the approach to new problems, the policy that emerges is influenced by what we can call the "shadow of the past."

Describing the character of large-scale political structures and understanding the impact of those structures on policy outcomes are two separate but related tasks. The first enterprise is better developed than the second. Scholars have accumulated a rich literature on what Charles Tilly has called "big structures, large processes, and huge comparisons." They have been primarily concerned with explaining broad-gauge similarities and variations in the emergence of national states since the sixteenth century and the transformation of industrial society in the nineteenth century and twentieth centuries. Scholarship of this sort is what Tilly calls "macrohistorical." He argues that "within a given world system, we can reasonably begin to make states, regional modes of production, associations, firms, manors, armies, and a wide variety of categories . . . our units of analysis. At this level, such large processes as proletarianization, urbanization, capital accumulation, statemaking, and bureaucratization lend themselves to effective analysis." The task of the analyst, according to Tilly, involves

[3]Arthur L. Stinchcombe, "Social Structure and Organizations," *Handbook of Organizations*, pp. 148–49.
[4]James G. March and Johan P. Olsen, "The New Institutionalism: Organizational Factors in Political Life," *American Political Science Review* 78 (September 1984), 741.

tracking down "uniformities and variations among these units, these processes, and combinations of the two."[5]

This macrohistorical enterprise frames the overarching structures and processes in which struggle individuals, groups, and classes. It gives analytic form to otherwise undifferentiated historical process, and it also helps reveal the historically contingent nature of particular political outcomes. At the same time, it has difficulties in developing causal explanations of political and social outcomes, and it does not provide a basis for understanding the microfoundations of large-scale structural continuity and change.[6] While drawing on the insights of the macrohistorical approach, I have sought to avoid such problems in two ways. First, I have cast the concept of organizational structure at a more proximate level, developing the idea of state structure in ways the relate directly to the limitations on and possibilities for policy action by executive officials. Second, I have focused more closely on the activities of executive officials as they interact in the policy process and adapt to prevailing organizational structures. In sum, I am concerned to explain not just the shape and variations of large-scale political and social structure but in how specific types of structure serve to constrain and channel the activities of politicians and administrators.

Accordingly, a focus on the organizational structure of the state is useful. State structure, understood as a piece of strategic terrain that influences the course of political struggle over policy, is important in several ways. First, the organizational structure of the state fixes in place the resources and policy tools available to executive officials and other actors in the policy process. The coherent planning ability of executive officials, the policy instruments available to government officials, and the bargaining resources at their command are established by government institutions. State structure, in other words, is a platform from which executive officials pursue their own policy agenda. Second, the organizational structure of the state also influences how societal groups are able to gain access to the policy process. Again, prior policy outcomes weigh heavily in shaping what societal groups perceive to be at stake in the policy process and in influencing their standing within that process.

The capacity of executive officials to plan and implement national policy hinges fundamentally on the broader historical course of politi-

[5]Charles Tilly, *Big Structures, Large Processes, Huge Comparisons* (New York: Russell Sage Foundation, 1984), pp. 63–64.

[6]A critique along these lines is offered by Ira Katznelson, "Rethinking the Silences of Social and Economic Policy," *Political Science Quarterly* 101 (Summer 1986), 313–17.

cal development and state building. The fragmented character of executive institutions and the limited scope of direct and selective U.S. government involvement in the economy, especially in the energy area, were shaped at particular historical junctures in the political and eonomic development of the United States. Institutional innovation and reform occurred at moments of crisis, particularly major wars. Unlike its European counterparts, however, the American state became involved in the economy and in the energy area after, rather than before, the establishment of large private enterprises. Also, the types of industrial conflicts that brought the state into the economy differed from those which unfolded in Europe. The state remained on the periphery of business enterprise, primarily as a peace keeper and a regulater.

At the same time, the administrative capacity of the state to plan and coordinate policy within and across policy areas remained, in comparative perspective, underdeveloped. In the energy area, during crises of national security when the role of the state was under review, powerful private groups, working through Congress, blocked the expansion of federal responsibilities. The sequence and timing of historical events thus defined the central features of American state building.

The politics of American energy adjustment in the 1970s was a struggle over the appropriate scope and powers of the state. The ability (and inability) of the state to expand its responsibilities was rooted in and mediated by an earlier history of state building. State institutions do not simply flex to the needs of the moment or to the wishes of executive officials who command those institutions. Crises in earlier decades had generated proposals, even institutional experiments aimed at developing new state capabilities, but the new arrangements did not take hold. The established arrangements of business and government limited the state to a minimalist role. National energy goals were thus pursued within a fragmentary institutional structure, one that conferred little coordinated planning capacity on executive policy makers and provided few instruments with which to shape private behavior.

AMERICAN ENERGY ADJUSTMENT IN THE 1970S

In responding to international economic disruptions, all states seek to fashion policies that will improve the economic and political standing of the nation. In responding to the oil shocks, executive officials in the industrial importing nations articulated broad adjustment goals. The range of adjustment strategies—international and domestic, offensive and defensive—are related to the capacities of states. Differences in international and domestic capabilities are crucial in understanding the strategies that states in fact pursued.

The more powerful a state is, in international terms, the more emphasis it will give—at least initially—to international strategies of adjustment. "International capabilities" (the state's access to and control over international regimes and its ability to influence the behavior of other states) are precisely the resources necessary to implement the international offensive sort of strategy that suits the interests of the powerful state. Similarly, a state that is weak internationally has fewer international options and will therefore emphasize domestic strategies. A similar logic holds for the domestic capabilities of the state. When the state is strong in relation to its economy and society, it has the option to pursue domestic offensive strategies. When the state is weak in relation to its domestic system, however, it will seek international strategies by which to respond to crises.

These observations suggest that it was differences in capabilities that led states to pursue initially divergent policies of energy adjustment in the 1970s. In 1973 the United States had few domestic options with which to pursue energy adjustment. Oil prices were controlled, which prevented the rapid adjustment of domestic consumption and production. Moreover, the state's standing in the energy sector was in comparative perspective quite modest. Capacities for the broad-scale planning and development of energy production and conservation were not available. By various measures, however, the international capabilities of the state were far greater than those of other industrial consuming nations. Because of its overall position, therefore, the state pursued an international offensive strategy. American officials, led by the State Department, attempted to bring the industrial nations together in schemes to roll back or to stabilize OPEC oil prices.

Broad circumstances converged to make the international response attractive to American executive officials. At a more proximate level, within the policy process, the same logic holds. Domestic options, such as oil price decontrol and the enlargement of state capacity to intervene in the energy sector, were politically costly and complicated, and initially they were beyond the reach of executive officials. The State Department agenda, on the other hand, held out the promise of less costly internal changes, and it could be conducted in virtual isolation from domestic politics. At various junctures, as the international strategy ran its course from 1973 to 1976, American diplomatic officials pledged to make domestic energy sacrifices. Indeed, the logic of international energy cooperation was based on mutual restraint as a means collectively to reduce the pressure on international petroleum markets. These pledges, for the most part, went unfulfilled. However, the strategy was pursued as a diplomatic initiative and so did not require executive officials to get involved in the tangle of domestic politics. The appeal of the international strategy can be seen both at the gross structural level and within the confines of the policy process.

In comparison to the United States, other industrial countries, particularly France, had greater opportunities to pursue domestic policies. Their international capacities were much smaller than those of the United States, their ability to alter domestic structures much greater. The proposition is that states which are not powerful internationally but strong domestically will have a dominant interest in domestic strategies of adjustment, particularly domestic offensive strategies. Although the European and Japanese governments differed in preferred type or combination of domestic policies, they pursued their policy choices at the expense of the international schemes proposed by the United States. The vigorous efforts of the French government to expand rapidly its civilian nuclear energy program and its consolidation of bilateral commercial deals with oil-producing countries exemplified the national approach.

What contributed to the fragmentation of relations among oil-consuming industrial nations? Conflicts, particularly between the American and French governments, were rooted in larger historical and foreign policy considerations. Yet our focus on divergent international and domestic capacities of the state reveals an important limit to the possibilities of cooperation. The international offensive strategy pursued by the United States was only one of an array of adjustment strategies in principle available. The value states attach to particular strategies hinges on considerations of state capacity, both international and domestic. When domestic options are foreclosed or politically costly, international options become more attractive. Conversely, when states have opportunities to adjust the economy and society to adverse international economic change, international cooperative schemes appear less valuable to them. As state capacities vary, therefore, international cooperation will be pursued with different levels of commitment. In the early moments of the 1973–74 oil crisis and later, when efforts were renewed, cooperation proved elusive.

The United States, as a result, was forced to fall back on domestic adjustment measures. Policy experimentation continued. The various strains of policy were not simply created when others failed—in one form or another all were present in the earliest moments of the energy crisis. Rather, executive officials gave changing emphasis to particular policies. An understanding of the sequence involved again turns on the character of domestic structure and the constraints it imposed on policies.

Even before the final defeat of the international schemes, American policy makers began to propose an expansion of the role of the federal government in the energy sector. Proposals both from within the executive branch and from Congress called for direct control of some phases of energy production and finance. They aimed to enhance state control over petroleum imports and consolidate national energy planning. The French had begun experimenting with a "national champion" in the oil industry back in the 1920s; for the United States in the 1970s, circumstances were similar. The political timing and the array of opponents, however, were very different. A powerful group of internationally integrated firms, under an American imprimatur, had emerged in the decades prior to 1973. Government agencies, despite crises and calls for reform, remained dependent on private firms for information and analysis. No body of trained officials—separate from industry and centrally lodged within state bureaucracies—had emerged in the United States. As a result, proposals were overwhelmed by interest groups and public officials in Congress and the executive branch. In the end, the fragmentation of scarce bureaucratic expertise and operational capabilities provided little on which to build new government powers. The eclipse of these proposals foreclosed efforts to expand the state's role in the energy sector.

Opportunities to establish state-owned energy entities had emerged in the United States at earlier periods in the growth of the petroleum industry. The failure, at earlier moments of crisis, to incorporate temporary mechanisms into the permanent policy apparatus of the state presaged failure in the 1970s. Some politicians and interest groups feared state encroachment in the private operation of energy finance and production. Others, including liberal Democrats, were skeptical of the purposes to which new state capabilities were to be put or concerned that industry interests would undermine them. The Rockefeller and Stevenson proposals did not have a legacy of successful state-run programs upon which to build. In the absence of precedents, and the constituencies that might have grown up around them, state-building proposals were easily defeated. The synfuels corporation, by contrast, had well-established congressional sponsors and a carefully circumscribed government role; its successful passage underscores the necessary ingredients for successful state building.

With the defeat of state-building proposals, executive officials next turned to traditional mechanisms. Most immediately, they spent money. The American state has always been well organized to spend money, which does not require much organizational infrastructure. Indeed, the limits to the fiscal instrument were more substantive than political or organizational. Programs reached their numerical limit by the late 1970s. More money could have been forced through existing channels, but the returns would have been unacceptably small.

The spending instrument was used in a way that conformed to the contours of American state structure. To the extent that a "state strategy" lay behind the disbursement of funds, it found programs that had technological merit particularly attractive, for two reasons. First, within the shifting environment of interest-group politics, the science and technology community was particularly suitable for organization by executive officials: the range of actors was more limited than in other policy areas, and executive officials could bring leading experts into the federal government to do the programmatic planning. Second, spending on science and technology can easily be presented as a "public good." It had few opponents, in Congress or within society as a whole. Money and science were brought together to build an extraordinary machine that would push funds into R & D projects.

The ease with which administrative officials could implement a massive spending program was not matched by immediate or substantial shifts in production and consumption. Oil imports continued to climb through the middle years of the 1970s, and increasingly the issue of oil pricing became prominent and disputed in national policy. As other policies either failed or produced few changes, executive officials came to embrace oil decontrol. Powerful societal interest had become attached to the maintenance of regulatory controls, however, and so the struggle for decontrol required an assertive government executive.

Other industrial consuming nations did not have regulated markets and did not confront this choice. For these nations the market was already working in 1973, and so the market was not an overt strategy. West Germany and Japan, for example, built on top of the market, using sophisticated mechanisms to encourage the competitive adjustment of their national economies to higher international prices for petroleum. The United States, on the other hand, had to remove layers of regulation to get to the market. First articulated by the Ford administration in 1975, this market strategy, after numerous concessions to Congress, culminated in the Carter administration's decision in 1979 to decontrol domestic oil prices.

Decontrol seems to avoid the goals or activities of the state. Deregulation may appear to involve the capitulation of public authority to private interests, in this case the interests of the oil industry, but the shift from controls to market is not one we can adequately explain in terms of the interests of private beneficiaries. Society-centered explanations for decontrol trace policy change to the changing capabilities of private groups. Such explanations are insufficient, however, because interest groups are highly mediated by the prevailing organizational structures of the state, and because changes in the capabilities of commercial and consumer groups cannot explain policy change. The interests of consumers, which otherwise would have been diffuse and difficult to organize, were crystallized in unanticipated ways by existing price controls. The dominance of oil industry interests before 1973 was undermined after the embargo by the strength of consumer interests represented in Congress. At the same time, the regulatory program fragmented the petroleum industry on the issue of decontrol. Some oil firms had short-term interests in the maintenance of controls. In other words, the regulatory structure and established institutional arrangements changed societal interests. The movement to market pricing for petroleum was not caused by the increasing influence of petroleum interests in the policy process. The decisive factor, instead, was the role of executive officials in marshaling resources to redefine the issue of market pricing.

A state-centered explanation highlights both the actions of executive officials and the institutional structure of the American state. Constrained by a fragmented and decentralized state structure, executive officials were still able to use the unique resources of the state. Of particular importance was the special access of state officials to the international system, manifest in 1978 at the Bonn summit where administration officials tied decontrol to a larger bargain on foreign economic policy. In turning decontrol into a foreign policy issue, executive officials called on special advantages over other actors who participated in the struggle to shape American oil policy. Policy change can thus be explained in terms of the initiatives of executive officials who, pursuing their own agenda, redefined the politics of oil pricing.

From the international offensive strategy of the Nixon State Department to the domestic decontrol strategy of the Carter administration, officials searched for workable policy within an institutional structure that provided few options. In 1973 and in 1979 the key government actors were foreign policy officials. In the first year State Department officials sought to use the powerful international position of the United States to implement a coordinated international response. In the last year foreign policy officials sought to use an international "bargain" to strengthen the executive in the domestic struggle over policy. In both cases, pursuing what was desirable and pursuing what was necessary, executive officials used a basic resource, one available to even a highly constrained state, namely access to the international system. When other options proved unavailing, moreover, state officials used their international resources to promote one of the most basic policy tools available to all states, namely the manipulation of the market.

THE ORGANIZATIONAL BASES OF STATE CAPACITY

This analysis provides a basis from which to advance our understanding of state capacity (the ability of a state to assert control over political and economic outcomes). States clearly differ in their abilities to make claims on national resources, to impose costs of international change on the economy and society, and to reshape the character of domestic institutions. State capacity has been termed "infrastructural power," by which Michael Mann means "the capacity of the state to actually penetrate civil society, and to implement logistically political decisions throughout the realm."[7] As I noted in chapter 2, the conceptualization of state capacity began in crude and ambiguous distinctions between "strong" and "weak" states. This book indicates that a more nuanced understanding of state capacity is needed, one that differentiates types of capacities within and among states and their requisite organizational foundations.

Distinctions between strong and weak states have been based on the manner and effectiveness with which states can intervene in the economy and society—differences that are in turn related to the centralization of "policy networks" and the policy instruments available to state officials. In schematic form the United States exhibits weak features: an organizationally decentralized and heterogeneous pri-

[7]Michael Mann, "The Autonomous Power of the State: Its Origins, Mechanisms and Results," *Archives européennes de sociologie* 25 (1984), 189.

vate sphere, as well as a fragmented and diffuse policy process. France is the counterpoint, with centralized and coherent bureaucracies, a wide range of policy instruments, and developed links with key industries and sectors.[8]

This literature is useful in identifying variations in the policy instruments and resources available to state officials. I have drawn on these structural aspects of policy making in tracing the efforts of executive officials to shape adjustment policy. Nonetheless, characterizations of weak and strong do not capture the elements of state capacity. The image of a weak American state—able to do little more than tally the demands of societal forces—is misleading, but it would be equally misleading to argue that I have provided evidence of a strong state. American executive officials, after all, were highly constrained in their initial search for a workable policy. It was the absence of domestic options that made coordinated international response to OPEC pricing so attractive to state officials. Yet those officials were not simply the captives of societal demands. The picture that emerges is one of officials searching for a proper "fit" between policy instruments and the demands of the energy crisis. In this sense, the capacity of the state must be discovered by those officials who seek to exercise it. Whether centralized bureaucratic states are stronger or weaker depends on the nature of the socioeconomic crisis at hand and on the possibilities of finding an effective fit between the available instruments and the particular problems. Moreover, the capacity of the state should not be measured solely in terms of the expansion of state controls or activities. The withdrawal of state control or abstension from intervention in the first place may also be evidence of state capacity.

Taken together, the literature on state strength is misleading in its basic implication that strong states are necessarily better equipped to assert power and control over economic and political outcomes. Three key propositions emerge from this book to illustrate its misleading nature.

First, strong states are tied to past policy commitments just as weak states are. In fact, a roughly inverse relationship exists between degree of intervention in the economy and society and degree of flexibility for the state. In a strong state, with many opportunities for

[8]See Peter J. Katzenstein, "Conclusion: Domestic Structures and Strategies of Foreign Economic Policy," in Katzenstein, *Between Power and Plenty*, p. 311, and Stephen D. Krasner, *Defending the National Interest* (Princeton: Princeton University Press, 1978), pp. 56–57.

intervention and tools to do so, subunits may themselves develop interests at variance with control at the top levels.[9] An analogue can be found in organization theory, with its negative correlation between bureaucratic rationality and the degree to which an organization is socially involved in its setting.[10] Policy tools and organizational resources that facilitate state intervention also create opportunities for social groups to thwart state policy goals.

It is with the development of state enterprises that ironies emerge. As one analyst observes, "the close relationship between state enterprise and government, deliberately constructed to ensure the precise operation of state policies can work both ways; state enterprises can use the apparatus to impose its ideas on government." In the French case, in particular, relations between the state and national firms changed: "State direction was clear and control was tight" in the years immediately after nationalization, but control later shifted.[11] Before 1973 the national oil industry sought to influence the energy priorities of government, to extend operational control over adjacent aspects of the industry, and to ensure financial access to state resources without political control by the state.[12] If the strength of the state lies in its control over political outcomes, simple measures of government involvement in the economy and society are misleading. The very act of intervention may prepare the ground for later subversion.[13]

Second, the state's withdrawal of regulatory involvement in the economy—"imposing the market"—may be as powerful an ex-

---

[9]See Peter Evans and Dietrich Rueschemeyer, "The State and Economic Transformation: Towards an Analysis of the Conditions Underlying Effective Intervention," in Evans, Rueschemeyer, and Theda Skocpol, eds., *Bringing the State Back In* (New York: Cambridge University Press, 1984).

[10]See, for example, Stanley H. Udy, Jr., "Administrative Rationality, Social Setting, and Organizational Development," in William W. Cooper et al, eds., *New Perspectives in Organizational Research* (New York: Wiley, 1964). The general point was made by a British civil servant quoted by Samuel H. Beer: "There is a general belief, as Government intervention expands, the power of the Government (Ministers and civil servants together) actually diminishes; that is to say the ratio between the responsibilities the government has taken on and its power to discharge those responsibilities become less favorable." Beer, *Britain against Itself: The Political Contradictions of Collectivism* (New York: Norton, 1982), p. 14.

[11]N. J. D. Lucas, "The Role of Institutional Relationships in French Energy Policy," *International Relations* 5 (November 1977), 120, 93.

[12]Ibid., p. 110. See also Harvey B. Feigenbaum, *Politics of Public Enterprise: Oil and the French State* (Princeton: Princeton University Press, 1985).

[13]This is the theme of Ezra Suleiman's major new study of the French state, *Private Power and Centralization in France: The Notaires and the State* (Princeton: Princeton University Press, 1988).

pression of state capacity as direct intervention. This use of the market as an instrument of state goals is an important counterpoint to direct intervention.

State capacity is difficult to gauge and compare. If a government intervenes in the economy to protect existing industry, either by imposing a tariff or providing subsidized loans, does its action indicate state strength? If a government withholds action, allowing a noncompetitive industry to decline, does it thereby indicate weakness? The problem with a priori judgments is that state capacities result in both action and inaction, in interventions, deliberate abstentions, and the withdrawal of interventions. The meaning of state capacity does not lie in the degree of direct activity by state organizations.

State intervention, though initially undertaken for national purposes, may provide a mechanism for private claims on state resources and thwart the original purposes of executive officials. Accordingly, the capacity of the state to extricate itself, or to resist intervention in the first place, is a crucial aspect of state capacity. This special kind of capacity—withdrawal—eventually became the central vehicle of American adjustment strategy.

Thus a fuller appreciation of state capacity must entertain the possibility that the imposition, extension, or maintenance of market processes in specific circumstances does not simply ratify the interests of societal actors but may be intimately associated with the state's pursuit of its own goals. Strategic abstention, withdrawal, and the reshaping of previous interventions are aspects of state capacity, just as much as strategic intervention—aspects we must appreciate analytically and be able to predict historically.

Third, the flexibility of state action—the ability of executive officials to provide themselves with the broadest array of options as they anticipate the next socioeconomic crisis—may be just as important as the degree of state control of the economy and society or the level of the organizational development of the state. The ability to intervene may be only a superficial indicator of strength if the state cannot also withdraw and redeploy resources.

The ability to generate independent preferences, extract resources from society, and shape private actions are integral to the capacities of states. Yet these abilities are not simply additive properties that allow us to locate different states along a single continuum of weak and strong. Rather, there are tensions between them: intervention may compromise autonomy; disengagement may enhance autonomy. Between countries, broad differences can be found in the institutional structure of state and society. Yet these differences are not only between states able to act and states overwhelmed by societal interests. The organizational structures of the American state are characterized by fragmentation and decentralization. Nonetheless, the American state does not simply register the demands of group interests. U.S. government officials have repeatedly found opportunities to assert state goals, particularly in the ability to spend money and regulate or deregulate markets.

The foregoing argument stresses flexibility as a crucial attribute of state capacity. The conclusion resonates with other nuggets of practical political advice; it emerges from the Machiavellian perspective that informs this book. I am concerned with the attributes of states, situated in domestic and international systems, that allow or prevent the realization of national goals embraced by executive officials. Robert O. Keohane's recent study of conflict and cooperation among advanced industrial countries arrived at the opposite conclusion: states might be better served by exchanging flexibility and "room for maneuver" for the predictability of international rules and procedures that structure the mutual expectations of governments.[14]

It is precisely the differences in the domestic capacities of states, I have argued, that lead to variations in the inclination and ability of governments to make international regime agreements. Divergent abilities to internalize the costs of international change had profound implications for the fate of cooperative proposals advanced by the United States in the 1970s. If all states had been equally incapable of adjusting their domestic economies and societies, of course, Keohane is right: cooperative agreements might have been achieved more readily.

The important political issue is whether governments should attempt to create new domestic arrangements to cushion the nation from the vagaries of change, or, alternatively, should push those problems into the international system and address them through regime agreements. Ultimately, the political stability of the international system might be strengthened by movement in either direction. Greater inflexibility of domestic systems would enhance the value of international regimes; flexible domestic institutions would place fewer demands on the international system. For the value of better means of internal adaptation we need only to examine the successes of many small, developed states in the international system. Such states have created innovative domestic arrangements that cushion

[14]Robert O. Keohane, *After Hegemony: Cooperation and Discord in the World Political Economy* (Princeton: Princeton University Press, 1984), pp. 257–59.

the nation and distribute the costs of fluctuations and shifts in the international economy.[15] If the major industrial nations are truly becoming "smaller" in relation to the international system, then the virtues of those domestic political arrangements may have broader applications.

[15]See Peter Katzenstein, *Small States in World Markets: Industrial Policy in Europe* (Ithaca: Cornell University Press, 1985).

Index

Cornell Studies in Political Economy

EDITED BY PETER J. KATZENSTEIN

Governments, Markets, and Growth: Financial Systems and the Politics of Industrial Change, by John Zysman

American Industry in International Competition: Government Policies and Corporate Strategies, edited by John Zysman and Laura Tyson

Library of Congress Cataloging-in-Publication Data

Ikenberry, G. John.
  Reasons of state : oil politics and the capacities of American government / G. John
Ikenberry.
  p. cm.—(Cornell studies in political economy)
Includes index.
ISBN 0-8014-2155-1. ISBN 0-8014-9488-5 (pbk.)
  1. Petroleum industry and trade—Government policy—United States. 2. Energy
policy—United States. I. Title. II. Series.
HD9566.I525 1988   333.79′0973—dc 19   88-3660